Th

Psychodrama
Papers

The Psychodrama Papers

John Nolte

ENCOUNTER PUBLICATIONS
97 Cumberland Street
Hartford, Connecticut

ISBN: 978-0-6151-9878-1

Contents

FOREWORD

About This Book

T his book is a collection of papers written over a period of about 35 years, and written for a variety of reasons. Some were written as handouts for a presentation at one or another psychodrama conferences. Some were written as handouts for a psychodrama training workshop. Several have been published. Others were written for discovery, attempts to articulate what I observed, knew or believed about psychodrama and the theory and philosophy upon which it stands. Each paper was written independently of the others and for a specific purpose. So this book is not an organized and integrated explication of the psychodramatic method. It is not a book on how to do psychodrama although it does present how I have dealt with certain issues that I have encountered as a psychodramatist, issues which I think other psychodramatists have or might also encounter. I expect it to be of interest mostly to practicing psychodramatists and students of psychodrama. I am publishing it because students of the National Psychodrama Training Center have found these articles helpful and I hope that others will find the same. I also hope that this material will engage the interest of psychotherapists of all persuasions and I will be delighted if reading this book encourages someone to explore the wonders and magic of the psychodramatic method.

About the Papers

Suzanne's Psychodrama is the protocol of a psychodrama session which I directed at the Moreno Institute in 1972. At that time, open or public psychodrama presentations were still being conducted at the Moreno Institute in Beacon, New York, and at the Moreno Institute in New York City. Moreno began the practice in 1942 in New York City and among others who attended one or more of the early sessions were Fritz Perls of Gestalt Therapy fame, and Sam Slavson who founded the rival group therapy association, the American Group Psychotherapy Association. In their travels in the 1950's and 1960's, the Morenos often demonstrated psychodrama by conducting a protagonist centered session.

I attended a number of these sessions in the New York City theater. Dr. Moreno would make some opening remarks, warm a member of the audience up to being protagonist and Zerka would then take over and direct a psychodrama. I was never completely comfortable with the extent to which people revealed themselves and explored personal problems in these sessions. Jonathan Moreno, in *Psychodrama since Moreno*, discusses the ethical challenges of the open session, focusing on whether the open session is theater, education, or psychotherapy. If it is the latter, he says, there are "intractable problems of confidentiality." Because of my discomfort with warming people up to personal problems and issues in an open session, I developed an approach to the open session which does not depend upon the protagonist exploring a personal problem. *Protagonist without a Problem*, included in this collection, describes this approach.

Catharsis, from Aristotle to Moreno is an explanation of the Morenean concept of catharsis which, he declared, was what was accomplished by psychodrama. Some confusion has developed because Sigmund Freud used the same term, catharsis, to describe the effects of Joseph Breuer's use of hypnotism with hysterical patients. Freud later soured on the cathartic method, finding that the results tended to be temporary rather than permanent. Today, catharsis, as described by Freud, and abreaction are considered controversial techniques. Moreno extended the term to cover not only the abreactive aspect of catharsis but an integrative aspect as well. It is the integrative catharsis which makes psychodrama effective.

I have used various versions of this paper to accompany presentations on catharsis at the Annual Conference of A.S.G.P.P. and at the conference of the Midwest Conference on Psychodrama.

Psychodrama and the Dimensions of Experience is another attempt to explicate how psychodrama has the impact that it has. Catharsis is re-visited. Then I try to explain some of the basic ideas in psychodrama. It is helpful to keep in mind that a psychodrama can be looked at as the externalization of the subjective. Our thoughts, perceptions, relationships and fantasies which are so subjective and ephemeral are made objective and tangible with the aid of psychodramatic techniques allowing us new ways of dealing with them.

Spontaneity–Creativity. Here I tackle Moreno's Canon of Creativity, the axiomatic principles of the creative process as laid down by Moreno. This paper reveals the sense I have made for myself out of Moreno's ideas. I have taken issue with him on a couple of points. For example, he maintained that spontaneity could be observed and I believe that like electricity, it is only the *effects* of it that can be observed. I have immense regret that Moreno died before I had reached this level of understanding of spontaneity and so did not have an opportunity to discuss it with him.

Having been well trained in a positivistic, objectivist approach to understanding the world I could not make sense of Moreno's notion of "unconservable energy," until I had read David Bohm's hypothesis of a subtle form of energy lying beyond detectability by today's instruments. Bohm's notion of the implicate order, described in *The Implicate Order* and other books, actually fits hand in glove with Moreno's Canon of Creativity. Explication of that is for another book.

Strategies of Directing came out of my attempt to understand differences in directorial styles and common elements among them. This paper was written for a psychodrama training workshop on the subject. I have presented this topic at an annual conference of the A.S.G.P.P. and in many psychodrama training workshops.

11

Psychodramatic Production of Dreams was co-authored by James Weistart and Jean Wyatt and published in *Group Psychotherapy and Psychodrama,* vol 28, p 70-76. I was on the faculty of Sangamon State University (now University of Illinois at Springfield). Jim and Jean were students in my course on psychodrama. Jean was the protagonist in a dream psychodrama and Jim had taken very careful notes during the drama. Then he, Jean and I reconstructed the drama as it is presented here.

Dream psychodramas seem to have a special quality which I have trouble describing and putting into words. I suspect it is because they arise from surplus reality, a term Moreno coined and which includes not only dreams but fantasies, delusions, and hallucinations. In any event, there is a different atmosphere in the theater when we are doing dream work. Moreno held that he purpose of dream psychodrama was not to interpret the dream but to put the dreamer in control of the dream. I thought that a reminder of dream psychodrama might be timely and the editors of *Group Psychotherapy and Psychodrama* agreed and published it.

Role Reversal with God is another article which came out of my psychodrama classes at Sangamon State University. It was co-authored by James Weistart and Cynthia Smallwood and published in *Group Psychotherapy, Psychodrama and Sociometry,* vol 30, p 37-38. Jim was the member of a mainline religion but one in which God had a lot of Old Testament features and was not all that approachable. His discussions with me about the drama gave me the idea for offering an article on role reversal with God to the psychodrama journal. I suggested that each of us write about the experience and then cut and pasted an article that is like a colloquy.

I have been aware of a predisposition on my part as a director to take any opportunity that a protagonist offers to do a role reversal with God. As in the drama presented here, this almost always provides a salutary effect for the protagonist. J. L. Moreno spent a few years of his young adulthood in a role reversal with God. Although this later laid him open to derision by members of his profession, he, too, found it a salutary experience and the inspiration for all his later work, including the Canon of Creativity, psychodrama, sociometry and group psychotherapy. He recommended God-playing to every one.

I am still deeply touched when I read Cindy's letter to her father.

Psychodrama in the Treatment of Incest and Other Kinds of Sexual Abuse . In 1975, I conducted a 3-day psychodrama demonstration workshop in Omaha, Nebraska . About 45 people attended. Early in that workshop the subject of adult-child sexual abuse emerged as a participant encountered her mother's boyfriend who had seduced her at age 14. I was taken aback by the number of women in the group who shared stories of sexual abuse, some at much earlier ages. This was about the period that the mental health community became aware of how widespread incest and the sexual abuse of children actually was. And still is. I first encountered an incest victim in 1960, early in my career as a psychologist. The reading that I found announced that only 1 in 10,000 women were subjected to incest! So much for statistics.

The psychodrama described in this paper was the last session of that workshop in Omaha. I think no one who was there has forgotten.

A Psychodramatic Perspective on Rage in Group Psychotherapy. In 1994 Ann Hale, then president of A.S.G.P.P. asked me to represent the A.S.G.P.P. at the Annual Conference of the A.G.P.A., the American Group Psychotherapy Association. My assignment was to offer a workshop and participate in a panel discussion of handling rage in group psychotherapy. I prepared the initial version of this paper as a handout for both. The psychodrama described was directed during that workshop and has been added to the current version printed here.

Protagonist Without a Problem. Other the years I have had a number of requests or opportunities to demonstrate psychodrama to groups where I felt that it was inappropriate to conduct a classical therapeutic psychodrama. Some of those groups were non-mental health professionals. One was composed of the members of a mental health association. Others were psychology or counseling classes. Building on J. L. Moreno's notion that every meaningful experience in one's life should be experienced twice, the second time in psychodrama, I started using the approach described in this short paper. I believe dramatizing a meaningful experience demonstrates that psychodrama is first of all an art form, and secondarily, a method of communication without asking potential protagonists to reveal life problems. Some of the most deeply touching psychodramas that I have directed have come from dramatizing a meaningful experience.

Role Training and Medical Education. I wrote this paper in 1975, and so it is dated. I wrote it for the head of a medical department who wanted to hire me to train the department physicians, interns and residents. He was unsuccessful but I have had opportunities, some of them described in the paper, to demonstrate how role training could be used in medical education.

A Brief History of Psychodrama at TLC. No, this TLC does not stand for Tender Loving Care. Here it refers to Gerry Spence's Trial Lawyers College,! In 1994, the year I turned 65, instead of retiring, I started what was essentially a new career by joining the staff of a new post-graduate trial lawyer training program, the Trial Lawyers College. Fellow psychodramatists Don Clarkson, Kätlin Larimer, and Kathie St. Clair share the credit for developing psychodrama from an introduction to personal growth and development à la the human potential movement of three decades ago, to being an integral element of the College. Even more credit goes to Gerry Spence who appreciated and recognized the power of psychodrama. He generously allowed and encouraged us to experiment with his college, to the benefit of both trial lawyers and psychodramatists.

I have long been an advocate for non-clinical applications of psychodrama and the other Morenean methods. Our experiences with TLC demonstrate how introducing spontaneity-creativity into a non-mental health setting stimulates, indeed unleashes, creative action on the part of those participating. In 1996, Ed Stapleton, a member of the charter class of the Trial Lawyers College from Brownsville, Texas,, and I discovered, almost by accident, how to use the psychodramatist as a trial consultant, and since that time Don, Kätlin, Kathie and I have continued to develop ways of utilizing psychodrama in this additionally new application of the method.

Remembering J. L. Moreno. The American Society for Group Psychotherapy, Psychodrama and Sociometry celebrated the 100th anniversary of J. L. Moreno's birth at the Annual Conference in 1989. President Sandra Garfield asked me to deliver a tribute to Dr. Moreno. This last section of this book contains my remarks.

A PSYCHODRAMA PROTOCOL: SUZANNE'S DRAMA

This drama took place one Saturday evening at the Moreno Institute in Beacon, New York, in the First Psychodrama Theater. At the time, in 1972, Saturday evening was an open theater night which means that anybody who wanted to could attend the session for a small fee. This particular night was during a training period, and perhaps a dozen students were present. Fifteen to twenty students from a nearby state college had come to see psychodrama at first hand.

The theater, once the carriage house of the summer home of a prominent family, was about 54 feet long and perhaps 24 feet wide. The ceiling was almost 20 feet high. Occupying about one third of the room was the stage, a round, three-tiered affair with a balcony some eight feet or so above the top circle and stretching across the entire back of the room. Two battens of colored lights were in the ceiling, lighting the stage and balcony and controllable by the director. The audience seating is simple, wooden folding chairs which are arranged in the rest of the theater. Approximately 70 people can be accommodated in the theater.

At 8:30, Zerka Moreno enters the theater and begins the session, welcoming the audience and then discussing the work of her husband and his contributions to the field of psychotherapy. She relates some antedotes from his life. The atmosphere is warm and friendly. The size of the theater increases the sense of intimacy that her manner encourages.

After perhaps half an hour, she asks to be excused. Ordinarily Zerka directs the Saturday evening open sessions but, she explains, she has an important trip to make early in the morning and she would like the rest of the evening to prepare for the morning. She has asked one of the advanced students, John, to direct for this evening and she introduces him to the group. She then leaves.

John takes the stage.

DIRECTOR: You know, there are people here tonight who have participated in dozens, even hundreds of psychodramas. And there are others who will be a part of their first psychodrama in the next few hours. Let's find out something about how much we, as a group, know about psychodrama, shall we? Who *has* been part of a psychodrama before?
(*Audience members raise their hands.*)

DIRECTOR: Looks like about half of the group are old hands tonight. The other question that I am interested in is this: Often people come back to the theater with the idea that perhaps they will have a psychodrama, that they will be the protagonist. I wonder how many of you have had that thought, have thought that you might want to be the protagonist sometime. Would you let me see. (*Perhaps a dozen people raise their hands.*)

DIRECTOR: Thank you. There! You see, we have no shortage of protagonists. We might as well steam ahead. Let me see your hands again. (*They go up again.*) Very good. Now, how many of you want to consider tonight as the night? (*Most hands quickly drop. A few go down more hesitantly. Finally, only Suzanne's hand remains up in the air. John knows Suzanne because she has been a member of the training group which started the day before. He also knows that she has trained as a dancer, knowledge that he will make use of later in the drama.*)

DIRECTOR: Come on up on the stage, Suzanne[1]. It looks as if everybody is willing for you to be the protagonist, but let's make sure, shall we? (*She nods.*) I'm sure some of the rest of you had some notion about being protagonist tonight. Remember that psychodrama is a group process, and it is important that the group have its choice of all possible protagonists. If you are interested in presenting something of your life tonight, don't be bashful at this point. (*There are no further volunteers.*)

DIRECTOR: I think that the group is saying that they would like for you to have your chance as protagonist. Shall we proceed? (*Suzanne nods.*) I would like to know if you have something in mind that you want to explore psychodramatically.

SUZANNE: Yes. My marriage.

---

[1]Names and other identifying information have been changed as is customary.

18

DIRECTOR: O.K. Tell me about your marriage.

SUZANNE: My husband and I lived together for two years before we were married, and everything went along just fine. Then we got married about a year ago and the relationship has deteriorated since then so that we have decided to separate for a while.

DIRECTOR: And you want to explore this because...?

SUZANNE: Because of all the pent-up feelings. I feel panic and nervousness. I feel upset. (*She shows signs of being anxious; her voice trembles a bit.*)

DIRECTOR: Just let yourself warm up to those feelings. Let them grow inside of you. Where do you feel this panic and nervousness?

SUZANNE: My throat gets tight like I'm going to cry, and my stomach churns, and my head pounds.

DIRECTOR: Do you feel very sad? (*SUZANNE nods.*) I want you to use your body to express what you are feeling. Get into the position that will best express what you are presently experiencing.

(*SUZANNE slumps to the floor, head down, back to the audience. She appears to be very sad and lonely.*
DIRECTOR dims the stage lights to a dark, deep blue.)

DIRECTOR: Are you very much in touch with your feelings now? (*SUZANNE nods.*) Would you give your feelings a sound?

SUZANNE (*Makes a whimpering cry*)

DIRECTOR: Again. (*Suzanne repeats the cry several times at the Director's instructions*)
What does that sound remind you of?

SUZANNE: Of a baby crying. (*Pauses*) I know. I'm a baby in the marriage relationship.

DIRECTOR: I want to know that feeling.

SUZANNE: Well, that's what Harvey always said.

DIRECTOR: That sounds like a frightened baby to me.

SUZANNE: I'm scared, yes.

DIRECTOR: How old are you, Baby?

SUZANNE: I don't know, I'm very young.

DIRECTOR: Just an infant, huh? About a year or less?

SUZANNE: (*nods*) Yes.

DIRECTOR: Where are you?

SUZANNE: In my crib.

DIRECTOR: Let's have a crib. (*A mattress is placed on the stage, and SUZANNE lies on it.*)
Why are you so scared, little baby?

SUZANNE: I don't know where my Mommy is. I'm all alone.

DIRECTOR: And what do you want to do? What do you feel like doing?

SUZANNE: I don't know.

DIRECTOR: Don't you want to call her?

SUZANNE: Yes!

DIRECTOR: Then go ahead and do it!

SUZANNE: Mommy! Mommy! Mommy! (*With the Director's encouragement, SUZANNE cries louder and louder. Her cry becomes a scream which seems to rise from a bottomless pit or despair and terror. It is obvious, a few minutes later when the lights are lifted, that the audience has felt the impact of this comment.*)

DIRECTOR: (*After SUZANNEf'S cries have subsided and she has sobbed quietly for a few moments.*)
SUZANNE, does this feeling remind you of some real-life experience in your past?

SUZANNE: Yes.

DIRECTOR: Let's go there, then. Where does this take place, and how old are you?

SUZANNE: I'm about 7. It's in St. Louis at home. I'm taking a bath. (*With Director's prompting, SUZANNE sets the scene. A mattress is used to represent the bathtub. Director asks who else is involved in the scene and SUZANNE mentions her mother and her younger sister Theresa who is three years old in this scene. Members of the audience are picked by SUZANNE, at the Director's request, to represent and enact the roles of Mother and Theresa. One is a student-in-residence, the other is a member of the audience whom SUZANNE has never before met.*)

DIRECTOR: Show us what happens. Do you take your bath by yourself? Does Mother help you?

SUZANNE: She gets me undressed and runs the water.

MOTHER (A.E.)[2]: (*Turning on water.*) It's time for your bath, dear. (*Mimes undressing SUZANNE.*)

DIRECTOR: (*To SUZANNE*) Is this right?

SUZANNE: Yes, except the faucets are at the other end.

DIRECTOR: O.K. Let's see how Mother does it. Reverse roles.

MOTHER (P): (*Turns on water*) Come on, SUZANNE, and take your bath. (*Undresses SUZANNE being enacted by Auxiliary Ego.*) Call me when you're ready to get out.
(*Director calls for a reversal back to original roles and scene is replayed. SUZANNE is now in the tub, taking her bath. She appears to be enjoying herself as she goes through the motions of playing in the water and washing herself. She is smiling and gives very much the impression of a seven-year-old playing in the tub.*)

DIRECTOR: Do you enjoy taking a bath? (*SUZANNE nods and smiles, almost sheepishly.*) Let yourself warm up to taking a bath at seven. Show us how you play.

---

[2]Parentheses indicate whether that role is being portrayed by the protagonist (P) or by an auxiliary ego. (A.E.) or a double ego (DBL.).

SUZANNE: *(Smiling.)* I do commercials.

DIRECTOR: What?

SUZANNE: I do commercials. I advertise things. *(Holds up imaginary bottle)* I can do an ad for Ivory Snow. Ladies and gentlemen, Ivory Snow is good to use. It makes bubbles. *(Pours from bottle)* See? I play with my toys. I splash. I splash with my toes.

Then I let the water out real sloowww. I feel the tub. It's warm and solid.

DIRECTOR: Now what happens. Do you call for Mommy?

SUZANNE: Yes. *(Confused)* No.

DIRECTOR: What happens?

SUZANNE: I hear something. A fight going on downstairs.

DIRECTOR: Who is it? Who's fighting?

SUZANNE: Mommy and Theresa. They always fight.

DIRECTOR: Where are they? *(SUZANNE points and says, "In the living room". Instructs Auxiliary Ego MOTHER and Auxiliary Ego THERESA to go to the area of the stage toward which SUZANNE pointed and MOTHER scolds THERESA. SUZANNE indicates that they are not doing it right. DIRECTOR directs her to reverse roles with MOTHER.)*

MOTHER (P): *(Suddenly screams in a rage)* I wish you were never born, you damn brat! I hate you! *(Etc., etc. Then she swings without warning, hitting THERESA on the arm.)*

DIRECTOR: Reverse roles! *(Scene is re-enacted with SUZANNE in role of THERESA, and Aux. Ego in role of SUZANNE. Aux. Ego is instructed to play MOTHER just as SUZANNE did, including hitting THERESA.)*

THERESA (P): Help! Suzanne! Help me!

DIRECTOR: Become yourself, Suzanne. Go back to the bathtub. Auxiliary Egos, re-enact the scene as Suzanne has shown you. Let's see what happens next.

(Scene between MOTHER and THERESA is re-enacted by Auxiliary Egos.)

22

(To Suzanne in the tub) What is going on inside you? What are you thinking?

SUZANNE: I don't know what to do. Theresa is calling for help and nobody else is going to help her. I don't want to go. Maybe Mother will hit me, too. I'm scared. I'm afraid. I don't know what to do.

DIRECTOR: What do you decide?

SUZANNE: I put on a towel and go downstairs. *(Goes through the actions.)* Don't hit Theresa. Stop! Stop it! *(continues pleading with MOTHER).*

MOTHER (A.E.): Get away or I'll hit you too! Stay out of this. It's between Theresa and me.
(SUZANNE indicates that she wants to play MOTHER'S role. Director calls for role reversal.)

MOTHER (P): *(screaming)* It's none of your business! This has nothing to do with you! I wish I had never had you damn brats! What did I ever have you children for? You kids are nothing but trouble! Get out of here!! Get OUT OF MY LIFE!!

DIRECTOR: Reverses roles and become Suzanne again. *(To Aux. Ego)* Repeat your last line. *(Aux. Ego as MOTHER does so.)*
(To SUZANNE) Now what happens? Do they stop fighting?

SUZANNE: Yes. They stop. My towel falls off and mother tells me to pick it up. *(Audience laughs. She shows towel dropping. Director calls for role reversal.)*

MOTHER (P): Oh, pick up your towel. *(Her voice has become normal, « and scene ends. Audience laughs in relief.)*

DIRECTOR: Now what happens next?

SUZANNE: I'm in the living room with mother. I'm crying and she is comforting me while I cry.
(The scene is set up, and SUZANNE is kneeling with her head in MOTHER'S lap. MOTHER is stroking SUZANNE's hair. SUZANNE indicates that something is not right about the scene, and Director calls for a role reversal.)

23

MOTHER (P): Suzanne, you are the only one who understands me. I don't know what I would do without you. You are the only one I like. I don't like either of your sisters.
(Roles are reversed and MOTHER (AE) repeats lines. SUZANNE is sobbing in MOTHER'S lap, but says nothing.)

DIRECTOR: Soliloquize, Suzanne. What are you thinking that you are not saying?

SUZANNE: (Soliloquy) Here I am again crying in the lap of this woman, my mother, when I really hate her. She is not a mother. She's awful. I wish I could trade her in for another model. I always prostitute myself to her to get Theresa off the hook.

DIRECTOR: Prostitute yourself? What do you mean, Suzanne?

SUZANNE: I mean that I sell my tears and comfort to get Mother's attention off Theresa.
DIRECTOR: This is the way you get Theresa off the hook?

SUZANNE: Yes. (Pushes away from MOTHER)

DIRECTOR: What happens next?

SUZANNE: I go to my room.

DIRECTOR: Fine. Let's see that. (He directs SUZANNE to set the stage as her room. The bed and other furniture are described, located, and chairs, tables, and other available stage furniture is used to represent that which was actually in her room. She is asked to look out the window and describe the scene, to describe the colors in her room, etc.) Is there some object in this room which is especially meaningful to you?

SUZANNE: Yes. Over here on this shelf is a tiny china doll that my Grandmother gave me. It was hers when she was a child.

DIRECTOR: Will you reverse roles with the doll? (SUZANNE strikes a pose.) I understand that you used to belong to Suzanne's grandmother but that you are now Suzanne's.

DOLL (P): Yes, that's right.

24

DIRECTOR: She treasures you very much? *(DOLL acknowledges.)* And you like her, too? *(DOLL agrees.)* Let's see now, how old is Suzanne right now?

DOLL (P): I guess she is about 11 now.

DIRECTOR: Is she happy? Does she have many friends?

DOLL (P): She's friendly but her family moves around a lot, so it is hard for her to keep her friends. They've been moving more than once a year for the last four years.

DIRECTOR: She doesn't have much of a chance to put down roots. It that it?

DOLL (P): That's right.

DIRECTOR: Is she sad much of the time?

DOLL (P): Sometimes she is. But she does have things she does that make her happy, too. She isn't always sad.

DIRECTOR: Thank you, Doll. Why don't you reverse roles and become Suzanne? *(To SUZANNE)* Would you put your doll back on the shelf and show us what happens when you come back upstairs after the fight?

SUZANNE: I rip up my bed and throw the sheets around. *(Goes through the motions.)*
(Soliloquizes) Where's Dad? Where is he in all of this? Why doesn't he control things, and fix things up? Why do I end up doing it?

(To DIRECTOR) Now Theresa comes up. *(He asks the appropriate Auxiliary Ego to come up on stage and calls for a role reversal since it is obvious that SUZANNE wants to re-experience an episode as it actually happened.)*

THERESA (P): Nobody likes me. I'm no good.

SUZANNE (AE): I like you, Theresa.

THERESA (P): I know. But Mom doesn't! She's a pig! I hate her!
(DIRECTOR calls for a role reversal)

25

SUZANNE: SShh! Don't say that. She's your mother! You should't say things like that! You should play up to her a little bit so that she'll like you. I do that. I get along O.K. with Mom.

DIRECTOR: Suzanne, what would you really like to say to Theresa, if only you could?

SUZANNE: Well, Theresa. To tell you the truth, I agree with you. Mom *is* a pig. I wish she weren't our mother. In fact, I envy you because you are the only one who is brave enough to stand up to that beast. You are doing the kind of thing that I should be doing, too. What are we going to do?

THERESA (AE): I don't know.

SUZANNE: Maybe we could all get together, we girls? *(Director calls for a role reversal.)*

THERESA (P): Maybe *(not convinced. DIRECTOR reverses roles again. SUZANNE indicates that this scene is over.)*

DIRECTOR: What happens now?

SUZANNE: After Theresa, Sandy, my other sister comes in.

DIRECTOR: How old is she?

SUZANNE: Three years older than me.

DIRECTOR: About 14 just now, eh? Would you pick someone out to be Sandy. *(SUZANNE does so, and DIRECTOR asks for a role reversal.)*

SANDY (P): God, I'm so depressed. Sometimes I think I'll commit suicide. Nobody understands me. I'm just so ugly. I wish I were dead. *(Roles are reversed, and the scene is re-played.)*

SANDY (AE): I'm so depressed, I think I'll just kill myself.

SUZANNE: Don't say that! It scares me. You're my only friend. *(Holds SANDY to comfort her. Then, to DIRECTOR)* I wouldn't be doing this. Not in my house. Nobody ever touches.

DIRECTOR: Except to hurt, is that right?

SUZANNE: That's it. Except sometimes Mom will comfort us if we are sick or crying. *(To SANDY)* I'm depressed, too, but I don't think about killing myself.

SANDY (AE): Nobody understands me. I'm just ugly. I wish I were dead.

SUZANNE: Don't talk like that, Sandy. Please don't. I don't want you to kill yourself. You are my only friend. *(Etc. Scene ends.)*

DIRECTOR: Well, does anybody else come to visit?

SUZANNE: Yeah. Dad comes in after he gets home from work. He knows something is up because Mom is crying downstairs. He comes up to ask me what happened.

DIRECTOR: You really hold court up here, don't you? *(SUZANNE agrees.)* Everybody comes to you for advice and comfort, huh?

SUZANNE: Yeah. *(DIRECTOR asks SUZANNE to pick an Aux. Ego. to play the role of Dad. She does so.)*

DAD (AE): Well, what happened today, Suzanne?

SUZANNE: Why don't you ask Mother? *(Role reversal)*

DAD (P): You know how she is when she's upset. Who started it? Which one? *(Role reversal)*

SUZANNE: I don't know, Dad. *(Auxiliary Ego stays in same line of questioning. SUZANNE indicates desire to reverse roles.)*

DAD (P): You know something? I have a theory. I wonder if these fights aren't sexual. I think that there may be something homosexual going on between them. They seem to have a love-hate kind of relationship going on between them. *(Role reversal)*

SUZANNE: *(stunned)* What?

DIRECTOR: Soliloquize, Suzanne .

SUZANNE: That never occurred to me. I feel weak, like something has shattered inside of me, like a glass crashing to the floor. I don't know what to think.

DIRECTOR: Reverse roles, again. *(She does.)* Dad , I would like to ask you a few questions if you don't mind.

DAD (P): Of course.

DIRECTOR: Do you really discuss matters of this nature with an 11-year old daughter?

DAD (P): Suzanne is really very mature for her age. I always discuss everything that is happening in the family with her. I usually find her comments are very insightful.

DIRECTOR: And you discuss a homosexual relationship between your wife and daughter with her.

DAD (P): Oh, I suppose it might not be the best thing to discuss some of this stuff with her.

DIRECTOR: Why do you think so?

DAD: Well, I know about the dangers of discussing sexual material with a daughter. They could be harmed by it, I guess. You know what I mean. But I really think she can handle it. She can handle almost anything.

DIRECTOR: You are a real disciple of Freud, eh?

DAD (P): Oh, yes.

DIRECTOR: Reverse roles, please. *(To SUZANNE)* What are your feelings about this man?

SUZANNE: I think that he is weak. He doesn't know how to run the family. I wish he would talk these things over with Mom. My sister *is* a lesbian, you know.

DIRECTOR: Wait a minute. How old are you?

SUZANNE: Oh, not when I'm eleven. She is today, though.

DIRECTOR: Alright. Fine. So there is your family. How do you feel about them?

SUZANNE: Ugh! They weigh me down. They are a burden on me.

28

DIRECTOR: Well, let's see that. Family, weigh Suzanne down. *(He gives them instructions to surround SUZANNE on the stage, and to move in on her, saying a key line. As they do, he has them each take hold of SUZANNE and bear down. They all speak together)*

MOTHER (AE): You are the only one that understands!

SANDY (AE): I'm going to commit suicide!

THERESA (AE): Help me$_s$ Suzanne!

DAD (AE): Suzanne, I need to talk some things over with you! *(As all the family presses on SUZANNE and bears down on her, she calls out for help.)*

SUZANNE: Hey, you guys. I'm trying to help you! The least you could do Is let me breathe! Help! Let me out of here! Help! Help! Help!

DIRECTOR: *(As family drag her to the floor.)* Who are you calling to help you, Suzanne? Anyone in particular?

SUZANNE: *(Pauses to think. Then with a sudden insight)* Harvey. *(Calls)* Harvey! Help me! Help me!

DIRECTOR: Harvey! Of course! Let's have a Harvey. Pick out a Harvey, Suzanne. *(SUZANNE selects a member of the audience to be Harvey.)* O.K., Harvey. To the rescue! Save Suzanne!

SUZANNE: Yes. Harvey, come on! Let's get out of here! Help me! *(HARVEY helps SUZANNE, pulling on her, and fighting the rest of the family loose. After considerable struggle, SUZANNE is pulled free and they run away from the other members of the family. SUZANNE is giggling and laughing, and hand-in-hand, they skip and run around the second level of the stage. SUZANNE is pulling HARVEY by the hand. During the struggle, the AUDIENCE is cheering them on, and gives a rousing cheer of victory when they emerge from the melee.)*

DIRECTOR: (After they pause for breath) Suzanne, you mentioned that the first two years that you and Harvey lived together, that you were very much in love and that things really went well for you together. Suppose you pick out a scene which would show us this happy time together for you and Harvey.

SUZANNE: *(thinks for just a brief moment)* I know. Sunday morning. Harvey always went out and got *a* big loaf of French bread, and we had a breakfast of French bread, cheese, and wine.

DIRECTOR: Will you set the scene. Where are we?

SUZANNE: Well, it is a typical student-type pad. It's kinda small and crummy. But I've got it fixed up really pretty nice. There is a big thing that a friend of mine painted hanging on the wall. Here is the bedroom in here *(Places mattress to represent bed.)* This is the living room, kitchen, and dining room combined. *(She shows how apartment is set up. The lights are arranged to give a reddish effect which she remembers as belonging to the apartment.)*

DIRECTOR: O.K. Let's see what happens. Who wakes up first in the morning? *(SUZANNE and HARVEY take their places on the bed.)*

SUZANNE: I do. *(Goes through the motions of waking up. Gets out of bed. Goes into other room. Moves around, straightening up.)* Then I do my exercises. *(She does the stretching exercises that only a trained dancer or acrobatic athlete could do.)* Then I wake Harvey up. *(Demonstrates.)*

DIRECTOR: Then what happens?

SUZANNE: I go back to bed. *(AUDIENCE laughs. SUZANNE blushes.)*

DIRECTOR: *(A bit embarrassed, too.)* Of course. Of course. And then what?

SUZANNE: We both get up, and he goes out to get the bread and cheese. I just bustle around and straighten up and get things ready.

DIRECTOR: Could you soliloquize your feelings, Suzanne.

SUZANNE: Oh, I'm feeling full of love. I feel free. I'm happy. It is the happiest time of my life. I feel alive and great.

DIRECTOR: Could you dance your feelings?

SUZANNE: *(Smiles. She likes this idea..)* Yes. *(For several minutes, SUZANNE dances. Her movements express an electric vitality, joy, tenderness, gentleness, and love. Her expression varies, but often a smile flashes, sometimes inviting or challenging, sometimes promising. Her body flows. When she finishes, many of the audience have been literally moved to tears.)*

DIRECTOR: That was beautiful. *(A few moments pass.)* Now here comes Harvey back with the bread and cheese. *(Harvey comes in and they prepare breakfast, opening and slicing the bread, uncorking a bottle of wine, unwrapping the cheese.)*

SUZANNE: Oh, you got Camembert!

HARVEY (AE): Yeah. I thought you'd like that.

SUZANNE: Oh, yes. Here, have some wine. *(Etc., etc. In a few minutes)* What are we going to do today?

DIRECTOR: Reverse roles.

HARVEY (P): Well, I've got to help Fred some this afternoon, and then we're going to a party at Jerry's this evening. *(Role reversal)*

SUZANNE: I don't want to go.

HARVEY (AE): Well, of course you want to go. And I want you to.

SUZANNE: No, I don't. All you are going to do is drink yourself into a stupor, or get high. I don't like that. I don't need it. *(Indicates that she wants to reverse roles.)*

HARVEY (P): The trouble with you, Suzanne is that you don't know how to have fun. You ought to try it sometime. Let yourself go. You got too many hang-ups. You think you're better than anybody else. Now you are going to go, and that's that.

DIRECTOR: *(Intruding into the conversation)* Excuse me, Harvey. But did you hear Suzanne say that she just didn't want to go? She doesn't care if you go, but she doesn't want to come along. Isn't that all right with you?

HARVEY (P): Oh, she'll come if I push her. She'll do anything I tell her to.

DIRECTOR: Oh. Why is that? *(Harvey doesn't answer)* Is it because she is afraid of losing you?

31

HARVEY (P): Yeah. She needs me. She doesn't think she can do without me. She really needs to learn to let down. Get stoned. She's too up-tight about things. *(There is a role reversal and SUZANNE capitulates, agreeing to go to the party. She gives the impression of being defeated, and of agreeing to go only to end the senseless argument which mars the delightful mood that was present j'ttst a few minutes ago.)*

SUZANNE: Then there is another scene just before we go to the party.

DIRECTOR: Well, let's see that.

SUZANNE: I come out of the bedroom wearing a sack dress, and Harvey says —

DIRECTOR: Let's see it. Reverse roles.
(SUZANNE (AE) comes out of the bedroom as per cue.)

HARVEY (P): What are you wearing that piece of shit for?

SUZANNE (AE): What do you mean? It's nice. I like it.

HARVEY (P): You look awful. You're thin and sexy, so you should dress like it. Wear clothes that show off your figure. Don't you care what you look like? *(Role reversal)*

SUZANNE: I care about me! Don't you understand? It's not what I wear that makes me beautiful, it's what I have inside! You can't push me around! You just want me to look good because you think I reflect on you like I'm your possession. I don't belong to you. I belong to me!

HARVEY (AE): I still say you look like a slob. Take that piece of shit off!

DIRECTOR: *(asks SUZANNE to reverse roles. Then says)* Harvey, do you really think she looks so bad?

HARVEY (P): Well, I think that she reflects on me. I don't want to be seen with a woman who doesn't know how to dress. I want someone I can show off. I like the prestige.

DIRECTOR: Well, what are you going to do if she won't change?

HARVEY (P): Oh, don't you worry. She'll change for me. I'll get her to do it in just a minute, wait and see. She needs my love too much so she'll change.

(To SUZANNE) O.K. Do what you want to do. You look like a pig in that thing, but if *you* want to look like a pig ... *(Shrugs)*

DIRECTOR: Reverse roles.

SUZANNE: *(Hurt)* Oh, for God's sake. I'll change. I can't stand this. *(Soliloquizes)* I feel angry. He has won and I don't understand it. He always wins. I always end up doing whatever he wants me to. I didn't even want to go to this damn party, and now I'm not only going, I'm wearing what he wants me to, and not what I want to. *(Flounces off into bedroom to change. Scene ends)*

DIRECTOR: Now, Suzanne, you've told us that this was in the happier days. Perhaps it wasn't all so happy, eh?

SUZANNE: Well, there was the other side.

DIRECTOR: In any event, you and Harvey did find some way to get married. I think that maybe we should take a look at that next. How does that come about?

SUZANNE: Well, I had moved out into my own apartment.

DIRECTOR: This is getting more complicated. Let's begin with the time at which *that* decision was made.

SUZANNE: We are in the apartment. It is evening. *(She and HARVEY are at the table. To HARVEY)* Are we going to get married? *(Indicates she wants to reverse roles)*

HARVEY (P): Look, I just don't know where I am on that. You know I love you. But I just don't think I'm ready to take that step. I'm still trying to find myself. I just don't know where I am with that. You understand?

SUZANNE: Then I'm moving out.

HARVEY (P): Is there somebody else? *(He is angry)*.

SUZANNE: Yes, I've been seeing somebody.

HARVEY (P): *(He is very angry and hurt)* O.K. Move out. I don't give a damn. Go screw anybody you want to. I'm just not ready and I don't know where my head is on that. *(Scene ends)*

DIRECTOR: Fine. Now, Suzanne, how much longer is it after this that you and Harvey decide to get married?

SUZANNE: It's just a few weeks. I moved out, but pretty soon Harvey is eating most of his meals with me and staying with me most of the time.

DIRECTOR: I see. You moved out and Harvey came along. Is that it?

SUZANNE: Almost.

DIRECTOR: Let's go to the time that you decide to marry.

SUZANNE: O.K. We are sitting in my apartment at a table. Harvey starts. *(Role reversal)*

HARVEY (P): Well, why don't we do it?

SUZANNE (AE): Do what?

HARVEY (P): Let's get married. *(Role reversal)*

SUZANNE: Wow, what a change! What prompts this?

HARVEY (AE): I don't know. Don't you still want to get married?

SUZANNE: Well, of course. But what's bringing all this about? Have your folks been talking to you? *(Wants to reverse roles.)*

HARVEY (P): Yeah, they've mentioned it a time or two. But what's the difference? I thought you wanted to marry me.

SUZANNE (AE): Oh, I do. It's just that this is such a change.

HARVEY (P): Want a ring? *(Role reversal)*

SUZANNE: Of course. But not a diamond. I want an old ring to cherish.

HARVEY (AE): Let's go get it. *(Scene ends)*

DIRECTOR: So that's the way it happened?

SUZANNE: *(nods)* Yes.

34

DIRECTOR: And this was just about a year ago. And since you and Harvey got married, things have gone downhill. Right? So now you've decided to separate. I think that we should next look at the moment in which you made that decision. When is it? Where are you?

SUZANNE: We are in our apartment, sitting at the table. I have made dinner. We have eaten, and I am cleaning up. *(SUZANNE sets scene and Auxiliary Ego takes his place, SUZANNE moves into action, her face set in irritation and frustration.)*

DIRECTOR: Soliloquize please. What are you thinking as you clean up?

SUZANNE: We never talk any more. We don't seem to have anything to say to each other. We eat dinner in silence. I clear this table and Harvey reads his paper. He never says anything but "un huh". I'm so tired of this. Where's the excitement we used to have? He just sits there! *(To HARVEY)* How was your day?

HARVEY: S'O.K.

SUZANNE: *(sits down)* Harv, what's wrong with us?

HARVEY (AE): I don't know. What makes you think there's something wrong?

SUZANNE: We never talk any more. You come home and sit down and eat and read your paper. It's just not like we used to be any more.

HARVEY (AE): *(Reading paper)* Uh huh.

SUZANNE: *(after a pause)* You want to know what I think?

HARVEY (AE): Huh uh.

SUZANNE: Well, here's what I think. I think you're ready for an affair. You think I hold you back from playing around with drugs and sex. You think I limit you, right? *(Role reversal)*

HARVEY (P): That's right. That's how I feel. I want to get out of here and live it up. You stop me. I didn't want to tell you that and hurt your feelings.

SUZANNE: Let's separate for a while and see how it feels.

HARVEY (AE): O.K. I agree.

SUZANNE: Boy, that was fast. I didn't think you'd agree so fast.

DIRECTOR: What are you thinking, Suzanne?

SUZANNE: He's going to lose me, the dope. *(Her emotion is a mixture of sadness and frustration.)*

DIRECTOR: Tell him.

SUZANNE: *(With anger)* You're going to lose me. You'll miss me. *(Scene ends)*

DIRECTOR: And so you decide to separate? What happens now?

SUZANNE: Well, he asked me to help him find another apartment. I told him to do it himself. *(She is indignant about his gall in asking her help in leaving her.)*

DIRECTOR: And he finally found one. When does he actually move out?

SUZANNE: While I am up here. When I get back, he is supposed to be gone.

DIRECTOR: I think that there is one more scene that we should do. Suzanne, in your mind's eye, where do you think Harvey is tonight, right now. What do you see him doing?

SUZANNE: Oh, he is in his new apartment with a bunch of his friends. They are getting high.

DIRECTOR: Would you set the scene?

(SUZANNE arranges the stage to represent HARVEY's new apartment. She sees him with three friends of his from the hospital and a nurse with whom she thinks he has been wanting to have an affair. She instructs the auxiliaries in their roles. All are sitting around smoking and listening to the Rolling Stones. Everybody is getting high. The lights are low.)

DIRECTOR: (Takes SUZANNE to one side of the stage, asks all Auxiliary Egos to go into action, and SUZANNE to watch. After a moment—) How do you feel watching this scene? *(HARVEY (AE) is getting very cozy with the nurse, who has been given the name CAROL)*

SUZANNE: *(She watches and it is obvious that she is bothered. She doesn't seem to be able to put her feelings into words)* This is crazy. *(DIRECTOR asks her to pick a Double Ego[3]. She does so, picking a student-in-residence with whom she has become close.)*

DIRECTOR: Soliloquize, expecially your feelings, Suzanne.

SUZANNE: I don't understand this. Why is this happening?

SUZANNE (DBL.): Did he give me up for this?

SUZANNE: Yes- that's right. Aren't I better than this? Doesn't he care any more than that for me?

SUZANNE (DBL.): This doesn't make any sense!

DIRECTOR: Reverse roles with Harvey. *(Suzanne does so.)* Harvey, what does all this mean?

HARVEY (P): I'm just doing what I want to. Suzanne is all uptight. She doesn't know how to let down and have a good time. So I'm doing it with people who know how. What's the big deal?

DIRECTOR: Well, Suzanne doens't quite understand where she fits into all this, and why you have left her for this kind of scene. Can you explain it to her?

HARVEY (P): Yes, I don't love her any more.

DIRECTOR: Tell her.

HARVEY (P): *(To SUZANNE)* I don't love you. I don't love you!

DIRECTOR: Reverse roles. *(Instructs HARVEY (AE) to repeat lines. He does so.)*

---

[3]The Double Ego is asked to stand beside the protagonist, tune into the protagonist's feelings and express anything the Double feels that the protagonist is not expressing.

SUZANNE: *(Begins to weep.)*

SUZANNE (DBL.): What's wrong with me?

SUZANNE: Yes! What's wrong with me? *(To HARVEY)* Why don't you love me? Why? Why?

SUZANNE (DBL.): Nobody loves me. I try to help but nobody loves me. *(Pause)* I feel all alone.

SUZANNE: (Wailing) I'm all alone. All alone. Always always alone! *(cries)*

DIRECTOR: Are there some more screams inside you, Suzanne? Go ahead. Let them out.

SUZANNE: *(Screaming and crying)* Oh no, no, no no! NO! NO! NO! NO NO NO NO! *(DOUBLE screams and crys with SUZANNE. Her screams increase and there is no mistaking the intense pain that SUZANNE feels about the situation she is in. After some minutes)* God help me! Help me! I can't stand it! I can't stand it! I am going to die! *(She slumps to the floor)*

SUZANNE (DBL.): I just can't cry any more.

DIRECTOR: What's happened to you?

SUZANNE: I'm dead.

DIRECTOR: You're dead. It's all finished.

SUZANNE: *(Matter of factly)* Yes. I got tired of everything.

DIRECTOR: It's all done. *(SUZANNE lies quietly. After a few moments)* How does it feel to be dead?

SUZANNE: Peaceful. Calm.

DIRECTOR: O.K. Let yourself feel it. *(After some moments)* I would like you to go up to the balcony. *(SUZANNE picks herself up and follows instructions. DOUBLE stays on stage floor.)* Well, what do you think? Take a good look at this scene. *(SUZANNE begins to laugh in a scornful sort of way.)* What are you laughing about? What's so funny?

SUZANNE: She is! That's ridiculous, to die for that! *(She points to HARVEY and his friends.)* She must be crazy. What a dumb thing to do.

DIRECTOR: Do you have any other feelings toward her? Take a good look at Suzanne. *(She does so, and looks pensive.)* Do you know how she died?

SUZANNE: She died of a broken heart. *(Sarcastically)* How romantic for this day. arid age. That' s dumb.

DIRECTOR: Is that all you can say? Did you ever have a broken heart? Do you have any idea how much this women has suffered?

SUZANNE: *(Her mood changes and she shows concern)* Yes, I guess I do. I want to help her. *(To DOUBLE)* Don't die. You still have a lot to live for. Get up. *(DOUBLE doesn't move.)*

DIRECTOR: Why don't you come down here and see if you can help her get up.

SUZANNE: *(Returns to stage level. Comforts SUZANNE (DBL.)* Come on. You can get up now. You've got a lot to offer. You can get up, dear. I'll help you. *(Lifts and pulls up on DOUBLE without effect.)* Please! *(To DIRECTOR)* She's too heavy. *(Tries tickling to which DOUBle does respond.*

DIRECTOR: No fair tickling.

SUZANNE: *(More strongly, almost angrily.)* Listen! You're Phi Beta Kappa! You've got a lot to get up for. You're pretty and you're nice, too. You've got to get to work. *(DOUBLE is still lying on the stage. Turns to DIRECTOR in some despair.)* Nothing's working.

DIRECTOR: Why don't you go back to the balcony and take another look. *(SUZANNE does so.)* What do you see?

SUZANNE: *(Studies her DOUBLE who is still lying on the stage.)* I see how helpless she looks. She's dead weight. I could lift her up but I could never keep her up. She's got to learn to stand on her own two feet.

DIRECTOR: Do you have any idea of what it is that is weighing her down?

39

SUZANNE: *(Pensively)* No-o.

DIRECTOR: Come on back down. I have an idea, and I would like to check it out with you. *(SUZANNE returns to stage.)* I think that maybe Suzanne's family is still producing some weight for her to carry around, and that maybe she doesn't need that. Is that a possibility?

SUZANNE: I suppose so.

DIRECTOR: Let's see if we can find out. *(Dismisses DOUBLE and asks for the Auxiliary Egos who have protrayed SUZANNE's family to return to the stage. SUZANNE is in the middle of the stage and family is surrounding her.)*

SUZANNE: Wait! I can't handle them all at one time!

DIRECOTR: O.K. Let's take them one at a time. Who's first?

MOTHER (AE): *(Wraps her arms around SUZANNE)* You're the only one who understands me, Suzanne. *(SUZANNE struggles to get loose. The first time, Auxiliary allows her to get free too easily. DIRECTOR points out that this is not the reality of the situation, and instructs the Auxiliary to hold on as tightly as she feels SUZANNE's MOTHER did. The struggle takes place again. This time, MOTHER smothers her more vigorously. The AUDIENCE begins to cheer SUZANNE on. SUZANNE finally pushes MOTHER off the stage and AUDIENCE gives her a victory cheer.)*

DIRECTOR: Alright, who's next?

SUZANNE: Dad, I guess. *(Auxiliary comes up on stage.)*

DIRECTOR: What kind of hold does he have on you?

SUZANNE: *(Thinking)* Dad's on my head. He's on my mind. He puts pressure on my head with all his "shoulds" and "shouldn'ts."

DIRECTOR: Let's see how he does that. Reverse roles. *(SUZANNE becomes DAD and demonstrates on Auxiliary how she experiences DAD's pressure on her head, using her hands. They are reversed to original roles.)*

DAD (AE): *(Squeezing SUZANNE's head with his hands)* Look what you did. You shouldn't have lived with a boy when you weren't married. And now look. You should have stayed with him and stuck it out. It's

your fault that you are separating. *(Continues in this vein. SUZANNE moves around stage as if to escape, but DAD stays right with her.)*
DIRECTOR: What do you want to do about Dad? Want to get him off your mind?

SUZANNE: Yes! *(To DAD)* You should be easy to get rid of!

DAD (AE): Ha! Wait and see. *(SUZANNE begins to struggle to get out from under DAD's grip on her head. The two are all over the stage, and during the course of the struggle, which is more violent than the one with MOTHER, they fall to the stage. DIRECTOR gives instructions on fair fighting. The AUDIENCE is cheering SUZANNE on: "Come on, Suzanne. You can do it!) Eventually SUZANNE gets hold of the edge of the stage and with this leverage is able to push DAD off the stage. The AUDIENCE gives her a victory cheer, and DIRECTOR raises her hand as if she were a prize fighter.)*

DIRECTOR: *(Gives SUZANNE a few minutes to catch her breath)* O.K. Who's next?

SUZANNE: Theresa. *(THERESA (AE) comes up on stage and confronts SUZANNE. SUZANNE looks at her for a moment, puzzled. Then to DIRECTOR)* I guess that I don't have anything to fight over with her.

DIRECTOR: Then do what you want to do.

SUZANNE: I guess I want to say "goodbye." (To THERESA) Theresa, goodbye. I love you very much. You are a beautiful person and you are brave. I believe in you more than you believe in you, you know. I wish you knew how wonderful you are, but I could say it a million times and you wouldn't really feel it. I'm always your friend, always, no matter what. I hope we can keep in touch.

THERESA (AE): Goodbye, Suzanne, I'm doing all right. I'm doing what I want to do. *(SUZANNE Embraces THERESA and gives her a kiss. She holds her for a few moments, then releases her. Auxiliary leaves the stage.)*

DIRECTOR: And now for Sandy.

SUZANNE: Sandy is actually still in my life as a sort of friend. We keep in touch and see each other from time to time. She has become a sort of conventional person, married and concerned with her home and all. I like her friendship, even though it is sort of superficial, and she doesn't really understand me, and I guess I don't really understand her.

41

DIRECTOR: Tell her what you want to say to her.

SUZANNE: I feel sort of sad when I see you. I don't think that you are terribly happy, but I don't think that there is anything that I can do about it. But I do like our friendship and want us to keep in touch. Maybe somehow we can get a little closer.

SANDY (AE): I want to stay in touch with you, too, SUZANNE. *(They shake hands, and SANDY leaves the stage.)*
DIRECTOR: I guess that leaves HARVEY, What are we going to do with him?

SUZANNE: I still want Harvey in my life. It's just that we need some distance from each other. My space is here and his is over there. Sometimes our spaces overlap and sometimes we are far apart.

DIRECTOR: Place him where you want him. *(SUZANNE places HARVEY at Stage Right, on the lower level. He is close enough so that she can stand at the edge of the top level and reach out and touch him.)*

SUZANNE: His circle is just as big as mine, but it extends out there beyond the stage. There are times when we can touch and share. *(She demonstrates by stretching her hand to HARVEY, who takes her hand.)*

HARVEY (AE): *(Pulls on SUZANNE)* Come on closer, Suzanne.

SUZANNE: . No! *(Pulls back.)* This is close enough. You're trying to push things too much.

HARVEY (AE): Oh, come on, Suzanne. I want you to be closer like you used to be. *(Although SUZANNE is apparently firm in stating that she wants to keep some distance, it is obvious that she is not struggling more than halfheartedly. Her lips are saying, "NO, NO," but her body language is saying, "Maybe.")*

DIRECTOR: *(Calls for the DOUBLE.)* I want you to reverse roles with this Suzanne. *(The DOUBLE replaces SUZANNE who stands back with the Director and watches as the above scene is re-enacted.)* What do you think, Suzanne?

SUZANNE: I think she has to struggle more than she realizes. It's not going to be as easy as I thought.

DIRECTOR: O.K. Reverse roles again. *(This time, when SUZANNE and HARVEY begin the struggle it is intensified. It is obvious that HARVEY intends to pull SUZANNE back into his "life-space", and she resists much more strongly than before. In danger of being pulled off her stage, DOUBLE grasps her around the waist and both pull back from HARVEY.)*

SUZANNE: *(To HARVEY, during struggle.)* Cut that out. You have your circle and I have mine!

HARVEY (AE): Come on, Suzanne. You can get a little closer. I want you to be closer.

SUZANNE: Play fair!

HARVEY (AE): I am playing fair. Just come on over here.

SUZANNE: No, you're not. Get your hands and arms out of my circle! *(Finally SUZANNE and SUZANNE (AE) pull away from HARVEY)*

DIRECTOR: Do you think that this is how it is going to be, Suzanne?

SUZANNE: Yes. I think so.

DIRECTOR: Then why don't we stop here. You've worked very hard and shared a tremendous amount of your life with us. I think that some of your friends in the audience would like a chance to share back with you.

SUZANNE: *(She relaxes and with a sigh is suddenly aware of the amount of effort she has expended.)* O.K.

Suzanne and the Director sit on the top level of the stage at front, and he coaches the audience in the method of psychodramatic sharing in which each member is encouraged to relate to the protagonist an experience, thought, event in his life, feelings, of which he was reminded by his participation in the protagonist's drama. Tonight, a number of people share their experiences of having been brought up in families where parents are inadequate as Suzanne's were, in some way or another. Others discuss strained relationships with brothers and sisters. A number of the audience have undergone experiences of divorce or break-ups with lovers. Frequent themes include disequilibrium in the relationship due to struggle over direction or control or life styles.

As Suzanne listens, a number of things are happening in her mind. Some of the scenes of the drama which have had a dream-like quality to her take on a new dimension, and she realizes for the first time the extent to which she has been involved in exploring her life experience, feeling emotions which had long been suppressed and articulating them, often for the first time. She feels quite drained but with the echoes of all the emotions she experienced swirling around in her mind.

As she listens to the sharing, she is aware that by her courageous exploration of her life experience she has made a significant impact upon so many members of the audience. Gradually she experiences a calmness and a sense of coming back from a fantastic journey to the group which has accompanied her on this journey.

CATHARSIS, FROM ARISTOTLE TO MORENO

Introduction

Over the years, the concept of emotional catharsis as a therapeutic factor in psychotherapy has stirred considerable controversy. Catharsis has been regarded by some as a major therapeutic vehicle and rejected by others as of little or no value in accomplishing therapeutic goals. Empirical studies aimed at evaluating its therapeutic efficacy, surveyed by Nichols and Zax (1977), have produced mixed results, and have even raised the possibility that catharsis can be counterproductive. This state of affairs is particularly interesting when one realizes that catharsis is not a precisely defined or described term. It is, rather, a heading under which are lumped, and often written off, a number of expressive emotional phenomena (Scheff, 1979). The controversy, therefore, may reflect the fact that participants are discussing different processes, concepts, or phenomena, rather than that there are conflicting findings regarding a single process, concept, or phenomena.

Catharsis is a concept of special interest to psychodramatists because J. L. Moreno (1972) emphasized the role of catharsis so strongly in describing the positive impact of psychodrama. He also claimed to have extended the notion of catharsis from its original Aristotelian description of the effects of Greek tragedy upon the spectators, to describing the effects of acting upon the actors, thus creating continuity between the aesthetic and therapeutic situations. It is therefore important for psychodramatists to understand Moreno's concept of catharsis thoroughly in order to defend its use, and in order to use it most effectively. Despite the fact that it has been discussed in the psychodramatic literature in recent years, (Kellerman, 1984; Blatner, 1985), there are still some subtleties with regard to the process of catharsis which neither Moreno nor other authors have explicated as clearly as might be done. The purpose of this paper is to do that.

Catharsis as Discharge of Affect

Sigmund Freud first introduced the term, "cathartic method" into psychiatry to describe Josef Breuer's discovery that when one of his patients recalled the event in which one of her myriad hysterical symptoms had initially occurred, something she could only do in hypnotic trance, the symptom disappeared. Freud was quite impressed when Breuer, his erstwhile friend, colleague, and collaborator, related how he had effectively brought about the cure of one of his patients, a young female suffering from hysterical neurosis. She later became renowned as "Anna O." in their *Studies on Hysteria*, (Breuer, J. and Freud, S., 1957) first published in 1895. Under hypnosis, Breuer discovered, Anna was able to recall the various events in which one of her many symptoms had occurred. When she had traced the experience of the symptom to its initial occurrence, a process in which she would express, for the first tine, the affect appropriate to the situation, affect which she had originally suppressed for various reasons, the symptom disappeared.

At the time that Breuer described his treatment of Anna O. to Freud, Freud was still intent on a university career in physiological research, and had not began the neurological consulting practice which was to lead to his eventual fame. Hence it was some six or seven years later, after he had entered private practice, that he first made use of "Breuer's technique of investigation under hypnosis" (p 48). *Studies on Hysteria* is an account of the two physicians' experiences in using the cathartic method, Breuer's with Anna O. and Freud's with four other cases, all diagnosed as hysteria. Freud, at this time, consistently speaks of the "cathartic method" as the overall approach, and of the expression of affect during the application of the method as "abreaction." Thus, what most therapists would call catharsis today, Freud would likely have referred to as abreaction.

At this time, Freud believed that affect was the residual of unexpressed instinctual excitation, (primarily sexual) and that it could have the toxic effect of creating hysterical symptoms when it had become "strangulated;," i.e., blocked from expression. Abreaction relieved the symptoms, he thought, because once the strangulated affect had been discharged it no longer produced its pathological effects.

His faith in abreaction and the cathartic method, which he had initially touted as bringing about a "complete cure" of hysteria, faded as he discovered that the "cures" that he was effecting tended to be temporary, and he found that most of his patients relapsed, with a return of the original symptoms, within a year or so after treatment ended. This disappointment may have played a significant role in Freud's rejection of his original hypothesis that the etiological factors in hysteria were to be found in the patient's life experience, and his shift toward a belief that "instinctual urges" played the major role in etiology.

Although Freud abandoned the cathartic method and abreaction as a therapeutic technique, not all of his psychoanalytic colleagues did. Among the theorists who were at one time associated with Freud, perhaps Wilhelm Reich is the one who remained most concerned with the importance of unexpressed affect, its physical expression in body tissue (body armor), and the therapeutic importance of "discharging" it. Today, organonomy, bio-energetics and radix, the current neo-reichian modalities, still retain this emphasis, utilizing techniques which most therapists would label catharsis.

Primal therapy, New Identity Process (sometimes referred to as "Casriel's scream therapy"), and Gestalt Therapy as conducted by some practitioners, are some of the better known modalities which also tend to include catharsis as part of their therapeutic armamentarium, if not the most important part of it.

The common emphasis in all these approaches tends to be upon the expression, "release," or "discharge" of emotion, anger, grief, and fear being the predominant ones. Although no therapist would admit to accepting Freud's original "hydraulic" conceptualization of emotion (Fritz Perls called it "Freud's excretory theory of emotion"), many still seem to regard emotion as something which is stored up inside the individual and has a pathological impact until it is emptied out. Either the "retention" of feelings or that which is inhibiting the "release" of them, some kind of blockage, (like Freud's "strangulation"), is seen as a legitimate and important aspect of therapy.

Catharsis also tends to imply that the emotional unloading is overdue, that the feelings expressed belong to a different time and place, and have somehow been stored up in the individual's mind, unconscious, or body. The individual expressing grief in response to loss, or anger in response to frustration is not generally considered to be in catharsis or abreaction.

Aristotle and Spectator Catharsis

Moreno's conceptualization of catharsis begins by referring to the original concept of emotional catharsis, proposed by Aristotle as an attempt to explain the impact of the Greek tragedy upon its audience. The tragedy, Aristotle wrote, in his *Poetica* results in a *Katharsis* (cleaning or purging) of fear and pity in the spectators. And this, as far as we know, is about all he wrote upon the subject. Scholars believe that there is a lost volume of Aristotle's work, and that in it, he may have elaborated upon the concept of Katharsis and tragedy.

Although nobody knows precisely what Aristotle had in mind, almost everybody has, at one time or another, been profoundly moved by watching a play, or perhaps a movie, an experience which is difficult to describe or explain, but one in which we leave the performance knowing that we have had a significant emotional experience. We are left with an unmistakable sense of having experienced something quite meaningful, perhaps of having encountered some basic truth of life. This is surely the phenomenon which Aristotle called *Katharsis*.

This experience is not limited, of course, to dramatic performance. It can occur as a response to the experience of any of the arts, music, dance, poetry, fiction, paintings, and sculpture. It is the experience which all artists pray that their work will have upon those who experience it. It is the experience that the beholder seeks. This is the phenomenon that John Dewey calls the "aesthetic experience," or the experience of art, and which he explicates in the early chapters of his work on the philosophy of art entitled *The Experience of Art*.

Every organism, he points out, human or otherwise, is constantly in interaction with its environment, striving to maintain a relatively precarious balance. At any moment, it may be exposed to some danger which its surroundings presents, and at every moment it is in need of some element which it must call upon its milieu to provide--air or food, for example:

> Life itself consists of phases in which the organism falls out of step with the marching surrounding things and then recovers unison with it--either through effort or through some happy chance. And, in a growing life, the recovery is never mere return to a prior state, for it is enriched by the state of disparity and resistance through which it has successfully passed... Life grows when a temporary falling out is a transition to a more extensive balance of the energies of the organism with those of the conditions under which it lives. (Dewey, 1934, p. 14)

A sense of order and of orderliness develops, Dewey hypothesizes, out of these repeated experiences of being out of and then re-establishing equilibrium with the environment. The organism seeks harmony in a "moving equilibrium," an active orderliness based upon tension. It is this sense of harmony, which occurs when the living being has first lost its integration with the environment and then recovers union with it, that provides the prototype of the aesthetic experience. In short, the aesthetic experience results in an emotional feeling based upon the perception of order and harmony in the universe. The artist, in creating a work of art, is attempting to communicate to the world his/her personal experience of order and harmony in the world. This is one side of aesthetic experience. When the work of art stimulates a similar experience in the spectator or listener, another aesthetic experience occurs. This writer believes that it was this phenomenon which Aristotle observed within the context of the Greek tragedy that gave rise to his notion of catharsis.

Moreno endorsed and confirmed Aristotle's description of the effect of drama upon the spectator. However, he also identified the role and importance that spontaneity plays in the experience of catharsis (Moreno, 1976), a factor that Aristotle had not taken in account. This is to say, the cathartic impact is due in large part to surprise, to the fact that the spectator does not know what to expect. Thus, the drama (or other work of art) has its most powerful cathartic effect during the spectator's first exposure. While one does experience additional catharsis during subsequent performances, the effect tends to become less and less pronounced with increased exposure, and eventually a work evokes no further catharsis. At that point, one may not want to see it again.

The spontaneous element is one determinate of the extent to which a drama is cathartic. Another is the degree to which the spectator can identify himself and his emotional reactions with those of the hero or other actors. The more closely that the situation and emotional reactions of the protagonist resemble situations and the emotions which the spectator has experienced, the greater the catharsis is apt to be for that spectator.

Moreno and the Actor's Catharsis

Moreno's experience with the experimental *Stegreiftheater*, in which the actors created and produced the drama simultaneously, focused his attention on the catharsis experienced by them and how it influenced their acting. He believed that because the actor embodies the role, rather than merely observing it, the cathartic impact of drama tends to be more powerful for actor than for the spectator. This is especially so, he thought, to the extent that the actor's part is "one for which he has an affinity;" that is, if the playwright has expressed private feelings of the actor more effectively than the actor has ever portrayed them in his personal life.

Thus Moreno distinguishes between "spectator catharsis" and "actor catharsis," both responses to drama. However, he writes, he had another source of inspiration beyond Aristotle's Poetica for the notion of the actor's catharsis. The concept of the actor's catharsis comes not only from drama but from the religions of the East as well:

> ...The other avenue led from the religions of the East and Near East. These religions held that a saint, in order to become a savior, had to make an effort; he had, first, to save himself. In other words, in the Greek situation the process of mental catharsis was conceived as being localized in the spectator--a passive catharsis. In the religious situation the process of catharsis was localized in the individual himself. This was an active catharsis....One might say that passive catharsis is here face to face with active catharsis; aesthetic catharsis with ethical catharsis. (Moreno, 1975)

Again the notion of spontaneity is introduced in that it implied is a significant change, even metamorphosis of the self.

Moreno has now extended the concept of catharsis considerably beyond Aristotle's conception of something taking place in the spectator of the tragedy. Catharsis is something that occurs in the actors as well, even more potently than in the spectator. And it occurs in life outside of the theater. More than by seeing roles enacted and emotions expressed by others, it is by acting out roles which touch one deeply, that involve feelings to which one relates most meaningfully, that one masters those feelings and those roles.

Moreno's preoccupation with the actor's catharsis was brought to fruition in psychodrama. Because the drama in psychodrama comes from the protagonist's own life, the protagonist is always ensured of having a "special affinity" for his part, actually his "parts" for he has a special interest in all the roles in his psychodrama. Of historical interest is the fact that in the initial period of the development of psychodrama, Moreno tended to conduct psychodramas with only a single patient and a staff of auxiliary egos, reflecting the overvaluation of the actor's catharsis as compared to spectator catharsis. It was by happenstance that he became aware that patients benefited significantly from observing and taking part in their fellow patients' psychodramas, re-confirming the validity of the concept of spectator catharsis and its therapeutic effectiveness.

Catharsis of Abreaction and Catharsis of Integration

In his earliest discussions of catharsis, Moreno described what he would later designate as catharsis of integration:

> He [the protagonist] takes his father, mother, sweethearts, delusions and hallucinations unto himself and the energies which he has invested in them, they return by actually living through the role of his father or his employer, his friends or his enemies; by reversing the roles with them he is already learning many things about them which life does not provide him. When he can be the persons he hallucinates, not only do they lose their power and magic spell over him but he gains their power for himself. His own self has an opportunity to find and reorganize itself, to put the elements together which may have been kept apart by insidious forces, to integrate them and to attain a sense of power and of relief, a "catharsis of integration" (in difference from a catharsis of abreaction.). (Moreno, 1953, p. 85)

The integrative catharsis is the ultimate goal of psychodrama, and it is probably the phenomenon which Aristotle noted in the spectators of the tragedy of his time, and which John Dewey called the aesthetic experience. It involves confronting and making peace with factors from both within and outside of oneself, factors and experiences which, when they occurred, threw the individual off balance and out of harmony with self and world. As integration is achieved, bit by bit, equilibrium is restored, and the protagonist feels newly empowered and a sense of harmony, freed from ghosts from the past and more at peace. Old psychological wounds and hurts may be healed, and the individual may be ready to forgive adversaries who once seemed to make life difficult.

Abreactive catharsis, from Moreno's perspective, is essentially catharsis as it is conventionally conceived in mental health circles, influenced by Freud's emphasis upon the expression of affect associated with an old and repressed memory, and typically involves exactly that process. The need for abreaction comes about as the result of a universal human experience, that it is sometimes necessary or advisable for us not to give expression to our feelings. From time to time, we find ourselves in situations in which intense emotions are aroused but which, for any of a number of reasons, are not expressed. This is especially frequent in childhood, in which, for instance, a child may be afraid to express anger toward a parent. For many families, there are emotions which family members are not allowed to express without the possibility of punishment. In some, it may be anger.

In others it may be fear, interpreted as cowardliness, or emotional pain, interpreted as weakness. Even in adolescence and adulthood, most of us find ourselves in relationships with family members, teachers, colleagues, supervisors, and others in which discretion often dictates that we not be completely spontaneous with our emotional responses. In other words, we learn to suppress if not repress, deny, or otherwise contain rather than express what we feel in many situations. Although therapeutic amelioration is not required for every instance of this sort, any event in which emotion is not appropriately expressed can become an emotional irritant in the future--and the focus of abreactive catharsis.

If there is a distinction between the way the psychodramatist and therapists from other modalities deal with abreaction (or "abreactive catharsis") it is that the psychodramatist is probably less willing to deal with emotions in a non-specific way. Thus, if a protagonist says, "I want to deal with my anger," the psychodrama director will concentrate on identifying the object, at whom or upon what, the anger is aimed, before getting around to the expression of the anger. More than likely, the director will want to establish a scene, complete with auxiliaries, in which the anger has some meaning. Then the protagonist will be encouraged or facilitated in expressing the anger, if it is appropriate, within the context of the scene, or in discovering what more important feeling the anger may be masking. In some neo-Reichian modalities, or in primal therapy, on the other hand, the client may be warmed up to the expression of emotions such as anger, fear, or sadness without identifying the object of the emotion or the context within which it arises.

Sound psychodramatic practice usually calls for the psychodramatist to continue the drama after abreaction has occurred. The goal is to achieve a catharsis of integration--which may well involve a further expression of feeling. One of the most common techniques is the ubiquitous role reversal. Thus, having expressed her intense anger and pain at her father whom she adored. but whom she experienced as distant, emotionally uninvolved, and rejecting of her during her childhood, the director may ask her to reverse roles with the father who is then called upon to justify his behavior toward and treatment of his daughter. Through this process, the protagonist may begin to clarify or modify perceptions of the significant other which make the behavior of the significant other more understandable, hence resolvable. At times, resolution comes about from a single drama; in other situations, many dramas may have to take place before resolution is fully achieved.

Catharsis and the "Conserve"

At this point, we want to turn our attention to what it is about catharsis that makes it therapeutic. There are probably a number of different ways in which we might be able to look at the impact of catharsis upon the organism, from a physiological or neurophysiological level to a philosophical one. Since we are concerned here with catharsis within the context of psychodrama and the theories of J. L. Moreno, it makes sense to explicate the effects of catharsis from that perspective.

According to Moreno's Cannon of Creativity (Moreno, 1953, pp. 39-48), spontaneity operating in the universe results in novelty, the emergence or creation of new things. Moreno introduced the concept of "conserve" as the class of the products of the creative process. It is the conserve that gives some consistency to a universe which is otherwise constantly changing. The products of human creativeness are further delineated as "cultural conserves," and include not only the obvious tools and instruments which the human race has developed, but also languages, alphabets, as well as mathematical and musical notations. These then are utilized in the creation of other conserves such as words, thoughts, laws, philosophies, histories, dramas, and novels as well as all the musical works that have ever been composed.

One class of cultural conserves of special interest is that of roles. Roles, a concept borrowed from theater, are the ways in which the self is manifested in social interaction. Roles always imply a relationship. The role of mother or father, imply the role of child, son, or daughter; the role of teacher imply the role of student, etc. Roles have both collective and individual components. Thus there is "the role of the mother" in any given society, a more abstract notion, and this role is concretely manifested by every mother in that society, each a little differently than any other. Every individual takes multiple roles: family roles, community roles, professional or work roles, recreational roles, etc. In the Morenean framework, the role is the tangible expression of the personality or self, and personality can be defined or thought of as one's "role repertoire." The learning of a role tends to follow a common process which begins with an act of perception, seeing the role or role relationship as portrayed by other members of one's society. The next step is usually that of "role-playing," a kind of rehearsal of the actions which the individual perceives as germane to the role. The final step is that of actually taking the role in one's society. Each of these steps demand some degree of creativity on the part of the individual.

While conserves are extremely beneficial to humankind, they also pose some problems, Moreno cautioned. On the one hand, they provide a handy, ready-made means for dealing with problems at hand. On the other hand, if a society becomes too enamored of the conserve, it may neglect to nurture creativity and the creative process. This can lead to serious problems which can eventually be manifested in social unrest of many kinds, and even social revolution. The full ramifications of this idea are far beyond the scope of the current discussion which must be limited primarily to the conserve of roles.

Any time an individual takes a role, his/her actions in the role may either be accomplished in a repetitive, mechanical, stereotyped manner, on the one hand, or the individual may infuse the role with spontaneity, manifesting it in a fresh or novel way. College instructors who read the same set of lectures year after year in their classes, and many people in bureaucratic positions are examples of the first approach. The teacher who recognizes that every one of the hundreds of students he/she encounters in his/her profession, and tries to modify his/her teacher's role to accommodate them all is an example of the latter. Thus conserves can be helpful instruments when the individual uses them as flexible tools, or they can become ruts and rigidities in the way one lives, substitutes for the spontaneous, creative life which Moreno envisioned.

A problem is that life as we live it is full of unexpected surprises, and few of us have the spontaneity to respond adequately to all of them:

> A change may take place at any time in the life-situation of an individual. A person may leave or a new person may enter his social atom, or he may be compelled to leave all members of his social atom behind and develop new relationships because he has migrated to a new country. A change may take place in his life-situation because of certain developments in his cultural atom. He may, for instance, aspire to a new role....Or he is taken by surprise by new

roles in his son or his wife which did not seem to exist in them before....Influences might threaten him from the economic, psychological and social networks around him....As long, however, as he is unable to summon the spontaneity necessary to meet the change, a disequilibrium will manifest itself which will find its greatest expression in his inter-personal and inter-role relationships...It is a peculiarity of these disequilibria that they have their reciprocal effects. They throw out of equilibrium other persons at the same time. The wider the range of disequilibrium, the greater becomes the need for catharsis. (Moreno, 1975, p 16)

We have a tendency to believe that things are the way in which we perceive them. For the most part, this works well, and especially with the physical world. If we lived otherwise, always questioning the accuracy of our perceptions, it would be a terrible and uncertain world, and we would probably all be psychotic. However, with respect to other people, our perceptions are not always so valid. First impressions can be biased and distorted as we know from the familiar notion of transference. Even when perceptions of others are, or have become, relatively accurate, others can change, sometimes with little warning and startling rapidity. In order for us to maintain emotional balance, our equilibrium, we are frequently called upon to alter our perceptions, both of others and perhaps of ourselves. These changes in perception are essentially what the catharsis of integration is typically comprised of.

Our perceptions are, in themselves, conserves, products of a creative process. A change in perception essentially means creating a new perception. Generally the perception which is changed is the perception of a significant other, of oneself, or the relationship or situation one is in. At times the perception is of the world at large, or of one's relationship to the world. From this perspective, catharsis can be though of almost literally as seeing things in a different way, "in a new light." These perceptions then alter or "de-conserve" some of our roles and role relationships, allowing for new and hopefully more effective interpersonal interactions.

An example, taken from an actual psychodrama, may help explicate the catharsis of integration. The protagonist, "Jack", is a man in his early forties, struggling to resolve an unhappy marriage. He has had an affair with a young woman, "Sarah", who is also married. This relationship has been quite meaningful and emotionally intense to both partners, and they have considered leaving their respective spouses to be together. The woman has decided to stay with her husband, although she and Jack remain on good terms. Jack is dealing with a number of issues. He still loves Sarah desperately but honors her decision to stay in her marriage. He experiences intense grief over the change in the relationship, and at times, feels quite angry at Sarah for her decision. The events have also convinced him that he needs to end his marriage so that he can seek a more suitable marriage partner, even if it is not Sarah.

As protagonist in a psychodrama, he states that he wants to "let go of Sarah," that he recognizes that he is "hanging on to her," or to their relationship, that this is unhealthy for him and stands in the way of getting on with his life. The director stages a psychodramatic encounter encouraging Jack to express all that he is feeling. Jack tells the auxiliary Sarah how much their relationship has meant to him, how he learned what it really means to love someone, and how painful it is to him to let go of it. Pain is apparent in his voice and demeanor as well as in the content of his words. And little flashes of anger intrude into the interaction. "Why do you have to stay with Roger?" he demands. "You know you'll never have with him what we have together!"

In role reversal, as Sarah, Jack reproduces the speeches he has heard from Sarah. "You know why I decided to stay with Roger," "she" says. "I realize that I haven't really given him a chance. And when I do, he comes through better than I expected. I made a commitment and I have to stay with it until I have a good reason not to!"

"Sarah" also tells Jack that she experiences him as demanding upon her, at times threatening to engulf her, and that that tends to make her want to pull away from him. Jack is puzzled by what she means. "I have accepted the fact that you are going to stay with Roger, at least for the foreseeable future," he says, "And I am not waiting for you to leave him. I am sure that there is somebody else out there that I can love as I have loved you. I want to find her."

The director asks him to re-enact a situation in which Jack thinks Sarah experienced him as demanding. Jack does so. It is a situation in which they have been getting along very easily together, somewhat rare since Sarah made her decision to stay with her husband. In a spontaneous moment, Jack gives her a warm embrace. Sarah pushes him away with some anger and says, "Please don't do that! I've asked you." Jack feels quite rejected, hurt, and angry.

"I can't believe that you really don't want any physical contact with me," he says to her. "I didn't ask you to go to bed with me. I don't have an erection. I just wanted a nice warm, friendly hug."

In role reversal, the director re-enacts the embrace. As Sarah, Jack pushes the auxiliary "Jack" away: "It doesn't feel good, Jack," "Sarah" says. "It feels like you are demanding something from me, and I feel smothered. Something is not right."

The director observes that in the role of Sarah, Jack's pushing away of the auxiliary "Jack" seems quite genuine. He intervenes. Keeping Jack in the role of Sarah, he demands, "Tell Jack what it is, Sarah. Don't you think you owe him something?"

"I don't know what it is," "Sarah" replies in some frustration. "I would tell him if I knew. I do love him, too, you know. But I have made a decision to stay with Roger. I don't know if it is right or wrong, but I've made it."

The director continues to work with Jack, still reversed into the role of Sarah, trying to identify the feelings. Jack, as "Sarah," does experience the feeling of Sarah wanting to push Jack away. "It is like he is trying to hang on me," "Sarah" says, "Or like he is trying to capture me, or something. I like Jack. I think I would like to be able to hug him—but something sure doesn't feel right about it." The director reverses Jack into his own role, and the auxiliary repeats Sarah's words. Jack suddenly is overwhelmed with tears. "I know what it is," he says to the director when he can speak. "Tell Sarah," asks the director.

"You are right," Jack says. "I wasn't aware of it until just now, but in a way, I have been making demands upon you. I have promised myself to get out of my marriage, and you know what a struggle that will be for me. There is a part of me that tells me how much easier it would be if you were waiting for me when I get the divorce. I guess that unconsciously I have been asking you to help me get free. I know that I have to do that for myself. I don't mean to put that on you. I am sorry." Again, he breaks down in tears.

"And all this pain?" the director asks.

"It's partly that I am sorry that I did that to Sarah. And it is partly that I am angry with myself for doing such a stupid thing. But mostly, it's that I know how I've been hanging on to Sarah, and I know that I can't do that anymore. I really do have to let go--let go of the fantasies of what we might have had together. Let go of everything."

The change in perception involves both self-perception and perception of the relationship between himself and Sarah. Jack has not been aware of the "demand" that he was placing on Sarah, and instead, considered himself to be accepting her decision to stay with her husband. As far as he was concerned, he thought himself rather nobly forging ahead with his life, honoring his relationship with Sarah by his acceptance of her decision. He was puzzled by what he considered Sarah's ambivalence, her friendliness one moment and her pushing him away the next. Experiencing his "demand" or dependence upon her in the psychodrama, in the role reversal, allowed him to correct his perceptions. This, in turn, made Sarah's behavior understandable, and Jack was able to indeed "turn loose" of his hopes that she would come to his rescue.

Fantasies are a particularly interesting kind of conserve. Because of our capacity for creativity, all of us are potentially capable of expressing ourselves in far more roles than is physically, psychologically, or socially possible for most of us to do, given the constrictions of life. Sometimes we desire roles which are biologically impossible. Men cannot be mothers, for example, and only a few of us have the potential for being great athletes. Often access to roles is quite limited. Only one individual can be president of the United Sates at a time. Sometimes we lack the intelligence, the physical beauty, the strength, the talent, or the spontaneity to become the professor, the movie star, the athlete, artist, or great psychotherapist that we would like to be. Sometimes the family, the social class, the race, or the country into which we are born means that access to roles for which we have both the potential and the desire is closed off to us.

Most of us have at least a little bit in common with James Thurber's Walter Mitty, fantasizing ourselves in roles, imagining ourselves performing deeds which we are not apt to actualize. Most of us manage in one way or another to come to terms with our fantasies. We find adequate means of self expression within the constraints which life imposes. None the less, some degree of disequilibrium as a result of not having access to roles which are important to us is fairly commonplace. Psychodrama can sometimes provide relief by affording us the opportunity to bring to fruition on the psychodrama stage some of those roles which, for one reason or another, cannot be fulfilled existentially. This typically provides a means for integration, for making available to the protagonist aspects which otherwise remain attached to the fantasy.

At the extreme, we discover those individuals whose life is so unsatisfying that they turn to their fantasy life in severely pathological ways. Psychodrama can provide relief through catharsis even for delusional individuals. Vivid examples of catharsis brought about in this manner can be found in such Morenean case studies, as The Psychodrama of Adolf Hitler, (Moreno, 1959) and A Case of Paranoia Treated by Psychodrama, (Moreno, 1969)

But of even greater importance to even more individuals, was Moreno's vision of the psychodrama and sociodrama stages as ways to achieve social catharsis, to bring about changes in the social order, changes in the ways in which society perceives itself and its members. Moreno was the first "radical therapist." He rejected out of hand Freud's notion that neurosis the price that we pay for civilization, and that we must learn to adjust to society. It makes no sense, Moreno proclaimed, to teach individuals to adjust to a society that is sick. We must, instead, teach them to change it, he declared.

The Relationship between Abreaction and Integration

At first glance, the concepts of abreaction and integration may appear to be rather disparate processes. We have suggested that abreaction involves the expression of previously unexpressed emotion and integration involves a change in one's perception. This writer has sometimes thought that Moreno's ideas would have been more understandable if he had used the term catharsis for the catharsis of abreaction and a different term for the catharsis of integration. This certainly would have been more in keeping with mainstream thinking. However, despite appearances, abreaction and integration tend to be intricately linked together. To understand this, we need to consider the nature of experience.

Although in the field of psychology, perception, emotion, thought, fantasy, memory, and behavior are often studied as if they were separate processes, this is not the way they are experienced in life. Rather, all of these various functions can be thought of as dimensions of experience. Experience, at any moment is multi-dimensional; that is, experience is comprised of several (or all) of these processes going on simultaneously, and interacting freely with each other. Thoughts give rise to emotions which bring back memories which may stimulate fantasies, which may influence the feelings.

Here is an example of how this entanglement of the dimensions of experience might play out in a typical psychodrama: the protagonist presents a problem, for example, "I am uncomfortable when my boss corrects my work." The director has the protagonist re-enact an instance of this. In the reenactment, a behavioral process, the director facilitates the protagonist in experiencing the discomfort fully, an emotional process, and at this point the director asks the protagonist to recall an earlier experience in which these feelings were predominant, a memory process. The protagonist may recall feeling this way when vis a vis a parent or a teacher in the past. Now the protagonist, with the help of the director, typically produces and re-enacts one or more scenes in which he/she felt unfairly treated by the indicated significant other, and in which the protagonist was unable to express his/her feelings adequately, perhaps because of fear of punishment or rejection by the significant other. The director encourages and facilitates the expression of feelings which the protagonist had originally been unable to express. This, in a word, is abreaction.

Once this has been achieved, the director will help the protagonist explore the relationship between protagonist and significant other. Very likely, a role reversal with the significant other is the first step in this process. Any of a number of psychodramatic events may follow. The protagonist may discover previously unrecognized motivations, feelings, and thoughts which make the actions of the significant other more understandable or even acceptable. Or perhaps the protagonist "evens things out" with the significant other by extracting some just punishment for the unfair treatment, and by doing so regains a sense of empowerment. This is an initial step in the catharsis of integration which may be completed by returning to the original scene of encounter with the current-day boss. Having dealt with the "unfinished business," the protagonist should now be freer to make a more valid assessment of the situation and relationship with the boss, and to take any appropriate actions. If all this is successfully brought about, the protagonist will ordinarily experience a great sense of relief, composed in part of a sense of rectification of an old injustice, in part of an altered perception of the current-day boss, and in part of a sense of direction for future behavior.

The relationship between abreaction and integration, most simply put, is that abreaction often a necessary prelude to integration. Very strong and inadequately experience and expressed emotion, associated with a perception, may make it impossible for one to relinquish that perception in favor of a newer and more accurate one. In this hypothetical outline of a psychodrama, emotion, memory, and perception are originally linked. Emotions from past experience have become a part of the protagonist's current perception of the boss. These are identified. In the scenes from the past, which are really the memory of past experiences, again there is the entanglement of perception and emotion. When the emotion is transformed through expression, a change in perception of the significant other from the past can be achieved, either because the protagonist sees the significant other in a new light, or, can balance the scales of justice, and can thereby see him/herself differently. In short, a perception of the protagonist has changed. In a sense, the protagonist's past experience has been changed, and now the director can test the impact of this change upon the current problematic relationship, that with the boss. It is likely that, on the psychodramatic stage, the boss will now be perceived differently. This difference in perception provides cues for the protagonist with respect to future behavior with respect to the boss.

Categories of Catharsis

In his numerous discussions of catharsis, only a few of which have been referred to here, Moreno established several sets of "categories" of catharsis. We have already mentioned actor's catharsis and spectator catharsis, as well as catharsis of abreaction and of integration, probably the ones most commonly referred to. Blatner (1985) mentions catharsis of inclusion and spiritual (cosmic, in Moreno's terms) catharsis in addition to abreaction and integration. In *Psychodrama, First Volume*, Moreno (1947) lists somatic catharsis, mental catharsis, individual, and group catharsis. Elsewhere he refers to the catharsis of laughter, and to partial versus total catharsis.

To understand Moreno, one must keep in mind that none of these various sets of categories are meant to capture the whole concept of catharsis. Each, rather, throws some light upon the meaning of catharsis to Moreno. Catharsis of inclusion involves a recognition of the impact and importance which one has upon and for one's various social atoms. Cosmic catharsis involves the profound sense of being connected intimately with the universe, of being, in a real sense a part of all that is. Somatic catharsis relates to Moreno's notion of psychosomatic roles, (Moreno, 1946, p 157-160) and might be considered to involve "deconservation" or retraining in this realm. Mental catharsis involves primarily psychodramatic roles. Individual catharsis differentiates from interpersonal and group catharsis, and, with respect to psychodrama, can be thought of as the cathartic impact of the drama upon the protagonist, and upon the spectators individually. At the same time, it should be recognized that the protagonist's catharsis is quite likely to change some of the protagonist's interpersonal relationships, and hence leads to interpersonal catharsis as well. Interpersonal catharsis more specifically refers to resolving the emotional tensions which may develop between two (or more) individuals and is what we try to achieve in marriage and family therapy. While each member of the group may experience some degree of spectator catharsis, group catharsis does not refer to the totality of these individual catharses. Rather group catharsis is the purging and integration of societal or cultural issues, and occurs both in and out of the psycho- or sociodramatic situations. The events following the assassinations of President Kennedy and Martin Luther King, Jr., or the events following the acquittal of the attackers of Rodney King reflect a group cathartic process. Moreno liked to address events such as these through sociodrama.

Conclusion

The purpose of this paper has been to explicate and clarify Moreno's concept of catharsis and, in the process, to make the psychodramatic process a little more understandable. When approached from the Morenean perspective, it is obvious that catharsis involves far more than the mere discharge of affect that the term often signifies to the non-psychodramatic therapist.

The discussion and analysis of catharsis and the dimensions of experience represents a phenomenological approach to the understanding of psychodrama and psychodramatic phenomena. The writer is currently working on a much fuller exposition of psychodrama from this perspective in the belief that psychodrama can be most completely understood and utilized when it is considered to be a phenomenological method of exploring personal and collective experience.

REFERENCES

Blatner, A. (1985) The dynamics of catharsis, *Journal of Group Psychotherapy, Psychodrama and Sociometry,* 37,   157-166

Breuer, J. & Freud, S. (1957) *Studies in hysteria.* New York: Basic Books

Dewey, J. 1934 *Art as experience.* New York: Capricorn Books, G.P. Putnam's Sons.

Kellerman, P. (1984) *Journal of Group Psychotherapy, Psychodrama and Sociometry,* 36

Moreno, J. L. 1946 *Psychodrama first volume.* Beacon, N.Y.: Beacon House, Inc.

Moreno, J. 1953 *Who shall survive? Foundations of sociometry, group psychotherapy and psychodrama.* Beacon,   N.Y.: Beacon House, Inc.

Moreno, J. L. 1972 Introduction to the fourth edition, *Psychodrama first volume.* Beacon, N.Y.: Beacon House, Inc.

Moreno, J.L. (1975). Mental catharsis and the psychodrama, *Group Psychotherapy and Psychodrama,* 28, 5-32

Moreno, J. (in collaboration with Moreno, Z.) (1959) *Psychodrama second olume: foundations of psychotherapy.*   Beacon, N.Y.: Beacon House, Inc.

Moreno, J. (in collaboration with Moreno, Z.) (1969) *Psychodrama third volume: action therapy and principles of   practice*

Nichols, M. P. & Zax, M. [1977]. *Catharsis in psychotherapy.* New York: Gardener

Scheff, T.J. *Catharsis in healing, ritual and drama.* Berkeley: University of California Press.

PSYCHODRAMA AND THE DIMENSIONS OF EXPERIENCE

Psychodrama evolved from an experimental form of theater originated by J. L. Moreno in Vienna in the 1920's. Developed initially for psychotherapeutic goals, psychodrama is far more than a form of psychotherapy or group psychotherapy as it is often designed. It is above all a form of drama, an art form. Like all art forms, it is a method of communication. It is also a way for one to examine one's life experiences, a profound kind of reflection, so to speak, a way to explore the contents of our inner lives, often making sense out of what initially seemed like nonsense.

The handful of formal studies which confirm the effectiveness of the method pale before the evidence presented by untold numbers of individuals who have experienced the power of psychodrama from the role of protagonist, and who know that it has helped them clarify their thinking, enabled them to make important life decisions, given them courage and inspiration to continue to struggle with life's problems, and helped them to know themselves in more profound ways.

The question which we wish to address here has to do with how psychodrama works. What accounts for its unique effectiveness? How does it change things for those who venture to be protagonists? Just what does it do, and how does it do it?

Catharsis and Restoration of Emotional Equilibrium

The function of the tragedy, Aristotle wrote, is to bring about a *katharsis* of pity and fear, and like affects. He says very little more, and there are several theories of how he thought that a cleansing or purging was brought about in the spectators of drama. Moreno (1940) suggested that he meant that "drama tends to purify the spectators by artistically exciting certain emotions which act as a kind of homeopathic relief from their own selfish passions." The notion of catharsis was introduced into psychiatry by Sigmund Freud and Joseph Breuer (Breuer & Feud, 1957). Breuer had introduced Freud to the phenomenon he had encountered with one of his patients who, when Breuer induced a hypnotic trance, would apparently relive an experience which had obviously been emotionally traumatic. Upon coming out of the trance, she would be calm and composed, and free of the neurotic symptoms for which Breuer was treating her. Freud tried Breuer's approach, which he called the "cathartic method" with some of his patients. Before long, he discarded it because he found the relief that was obtained was often temporary. He also disliked the process of inducing hypnosis and discarded it in favor of free association.

For Freud, emotion represented the residuals of unexpressed "instincts," primarily sexual drives. Catharsis, then, represented the release of this excess energy. His ideas of emotion and catharsis have been characterized derisively by Fritz Perls as the "excremental theoryl/ of emotions, bringing to mind the use of the term, cathartic, in medicine as a synonym for laxative.

After giving up the cathartic method, Freud introduced the term "abreaction" into psychiatry. This refers to the experience and expression of affect associated withe recall of a repressed event, usually one which is traumatic. Although the "catharsis" has remained in the psychiatric lexicon, it seems to be used interchangeably with "abreaction," and few writers seem to make a clear distinction between the two processes. Any strong expression of affect, especially if expression of the feelings is ego-alien to the individual, is apt to be identified as catharsis. Actually, only psychodrama and perhaps some of the hypnotherapeutic techniques involve the "re-living" of a past event, the situation to which Freud first applied the term.

The emotions which are expressed in catharsis are typically considered to have been associated with previous experiences, and are often thought of as having been "stored up" inside the individual somehow, and as "released" by or through catharsis. Although few therapists or theoriests consider emotion to be the waste product of unexpressed instinct, there is still a kind of thinking which seems analogous to constipation common to those who champion the use of catharsis.

Moreno developed some fairly elaborate ideas about the process of catharsis with respect to psychodrama, extending the concept considerably beyond the meaning which Freud had given to it or which it has in psychiatry in general. He noted that catharsis was originally a term associated with drama, implying that, because of his involvement in drama and his development of a dramatic form of therapy, he was more entitled to define the concept in psychiatry than was Freud or presumably anybody else!

As indicated above, spectators of the Greek tragedy, Aristotle thought, were purged of certain untoward emotions, fear and pity in particular are named, by seeing those feelings being enacted upon the stage. The Greek root of catharsis has to do with cleansing. Moreno agreed with Aristotle in his understanding of the impact of drama upon audiences. However, he also thought that there was another aspect to catharsis. He thought that Aristotle had touched upon only a part of the process, the spectator's catharsis. This, Moreno held, was the passive form of catharsis, accomplished simply by the act of watching and experiencing the drama vicariously. Moreno maintained, however ,that there is another form of catharsis, an active kind of catharsis, the idea of which comes not only from drama, but from the religions of the East:

> . . . The other avenue led from the religions of the East and Near East. These religions held that a saint, in order to become a savior, had to make an effort;he had, first, to save himself. In other words, in the Greek situation theprocess of mental catharsis was conceived as being localized in the spectator--apassive catharsis. In the religious situation the process of catharsis was localizedin the individual himself. This was an active catharsis....One might say thatpassive catharsis is here face to face with active catharsis; aesthetic catharsis with ethical catharsis. (Moreno, 1940)

That is to say, the cathartic impact of drama may be even greater for actor than for the spectator, particularly if the actor's part is "one for which he has an affinity," if the playwright has expressed private feelings of the actormore effectively than the actor has ever portrayed them.

More than by seeing them enacted, it is by acting out our deepest feelings and fantasies that we are freed of them, Moreno claimed, and this process is the actor's catharsis. Since the drama in psychodrama is *always* the protagonist's drama, the protagonist can be ensured of having a "special affinity" for his part.

Aristotle, Freud, and most other therapists and theorists who find value in catharsis, consider catharsis primarily with respect to the expression or "purging" of feelings or affect. Moreno writes about the effect of psychodrama as "restoring emotional disequilibrium" and "increasing the spontaneity"of the protagonist. Life, he suggests, is full of uncertainty and surprises:

A change may take place at any time in the life-situation of an individual. A person may leave or a new person may enter his social atom, or he may be compelled to leave all members of his social atom behind and develop new relationships because he has migrated to a new country . A change may take place in his life-situation because of certain developments in his cultural atom. He may, for instance, aspire to a new role....Or he is taken by surprise by new roles in his son or his wife which did not seem to exist in them before....Influences might threaten him from the economic, psychological and social networks around him. (Moreno, 1940)

Furthermore:

> Excepting rare instances, therefore, but few undertakings of any of us ever get so much as started. Every one of us has ideas--"dreams"--of himself in a variety of situations. These we call "roles." Most of our roles remain in the "dream" stage--they are never attempted or begun, and any attempts at actualiz-ing our roles (rare as they are) remain, like most of our relationships, fragmen-tary , inconclusive, loose ends.
>
> The number of major and minor disequilibria rising from instances such as these is so large that even someone with superhuman moral resources might well be confused and at a loss. (Moreno, 1940)

These unexpected and unplanned for events call upon the individual to meet them through the individual's spontaneity. Most of us lack the spontaneity to deal effectively and creatively with the all of the unexpected events with which we meet in a lifetime. As our hopes, plans, and expectations get frustrated, we are thrown off balance emotionally. As we struggle to right ourselves and get our lives back on track, we are frequently confronted with even more uncertainty and surprise. Along comes another unexpected event, and we are buffeted from a different direction. Eventually, the accumulative disequilibrium may reduce our effectiveness in coping with life's demands to the point that we need some kind of assistance in regaining our balance. In order to cope satisfactorily with life, with the "slings and arrows of outrageous fortune," we must learn to be more spontaneous. Spontaneity, which Moreno (1953) defined as "an adequate response to a new situation, or a new response to an old situation," is conceptualized as a human resource which is expandable through training. He considered his psycho dramatic method to be the training medium *par excellence* for increasing spontaneity. For some, the world is too complicated and the demands made upon them by the groups in which they live is greater than their resources or desires. For others, the desire to develop and realize more roles than are available predominates. For all of these situations, Moreno called for catharsis such as psychodrama provided.

Catharsis, in Moreno's sense, "deconserves" the individual, helps free one from the ruts and cliches that we so easily fall into, and in so doing results in the greater spontaneity which is necessary if one is to deal adequately with all the uncertainties which most of us must confront as we live our lives. "Life is what happens to you while you are busy planning something else," John Lennon is reputed to have said. This was very much Moreno's point of view, and spontaneity is the neglected human resource which he thought could get us by.

In addition to thinking of catharsis as "active" or "passive," Moreno made another distinction between the catharsis of abreaction and the catharsis of integration. The former refers to emotional expressiveness; the latter to a reorganization of self perception which can occur and may need to occur after the protagonist has recognized and expressed heretofore unexpressed feelings and roles. This additional aspect to the notion of catharsis is probably that which most markedly distinguishes the psychodramatic concept of catharsis from catharsis as it is conceptualized in most of the other modalities of therapy in which it is valued and utilized.

The catharsis of integration can come about in the following manner. During the course of a psychodrama, a protagonist may come to a realization that he/she has been unfairly treated, manipulated or intimidated by someone, perhaps a parent, a teacher, or a boss, and that intense resentment and anger, which he/she was afraid to express at the time, can still be aroused by recalling or re-enacting the event. The psychodrama director will ordinarily encourage the protagonist to express those feelings within the security of the psychodramatic situation. That is, the expression of the feelings within the psychodrama will not result in abandonment by the parent, being flunked by the teacher, or fired by the boss. Once this has been satisfactorily accomplished, the psychodrama director will additionally encourage the protagonist to experience the event from the role of the significant other. In the course of doing this, it often becomes apparent to the protagonist that the significant other was unaware, for one reason or another, of the injustice or the extent of it that was done to the protagonist. Sometimes it becomes apparent to the protagonist, now in the role of the significant other, that there were reasons for the actions of the significant other that the protagonist did not or could not take into account in evaluating the actions of the significant other from his/her own role. Or, from the different perspective, the protagonist may re-evaluate his/her responsibilities for the course that the situation took. For example, the protagonist may decide that he/she was afraid to stand up for him/herself adequately at the time of the original event.

The effect of this further exploration of a problematic event is that protagonists are often able to restructure their perceptions of such events, to see them in a new light, to re-evaluate the events in such a way that anger and resentment are no longer necessary or appropriate feelings for memory of that situation to arouse. When this happens, resolution has then been achieved and being reminded of the event no longer stirs up the old feelings. It is this process of resolution and relinquishing the need to maintain the strong negative emotion to which Moreno referred as the catharsis of integration and which he saw as the ultimate goal, more important than simply expressing the experienced but unexpressed feelings.

Finally, Moreno also made a distinction between "personal" and "interpersonal" catharsis. The vast majority of "disequilibrium" is generated in interpersonal relationships. Although one must frequently attempt to restorebalance without the presence of the other person, and while this is certainlypossible, "it is impossible to find the true seat of the disequilibrium; bothpeople are necessary, and they must be brought together in a situation whichis crucial for them and in which they can act spontaneously." The psychodrama stage provides a unique venue for this activity .

Having briefly examined the concept of catharsis, we are now ready totackle the issue of the relationship between spontaneity and catharsis. Actually, there are at least two questions to be confronted: How does catharsis increase spontaneity? And, what has spontaneity to do with the process of psychotherapy and emotional well-being?

The concept of spontaneity and problems of dealing with it have beenintroduced in an earlier chapter on Morenean philosophy, and will be con-sidered again in considerable detail in the upcoming chapter in which therelationship between spontaneity and emotion will be considered. For now, thestrategy for understanding the effectiveness of psychodrama will consist ofdescribing a psychodrama session and analyzing it in terms of the protagonist's experience. Changes in the way in which the protagonist is feeling about herlife and experiencing herself at the beginning of the psychodrama and how shefeels after the drama will be noted and commented upon.

A Typical Psychodrama

A common pattern among psychodramas begins with the protagonist presenting a troublesome emotional reaction that the protagonist is experiencingin present day life. The following is an actual example. A woman in atraining group states that she would like some help in making an importantdecision which confronts her. It seems to her that she is in a no-win situationand will have regrets regardless of how she decides the question. She also lets the group know that she is feeling very badly about herself, and it appears toher as if she has made nothing but mistakes her whole life. She has no confidence that she can do anything right. The group selects her as the protagonist for the session.

The director recognizes thather prevailing mood, her feeling that she has botched everything that she has ever done in her life is a considerable overreaction, and chooses to focus upon that aspect after first learning that the decision she must make has to do with continuing a lawsuit against the individual who involved her in an auto accident which had both material and emotional consequences for herself and her teenage daughter. She asks the protagonist to identify some of her many failures. The protagonist proclaims that she has failed as a mother (her daughter wants her to teach her how to drive, and she can't bring herself to do this), in her profession (her supervisor has criticized her for acknowledging, truthfully,to a patient that she doesn't have the answer to the patient's question), as a wife (she was married for a number of years to an emotionally and physically abusive man who sexually abused the daughter), as a lover (her current relationship is experiencing considerable tension), and so forth.

Part of the damages that she is suing for are emotional damages to the daughter who hadjust gotten a learner's permit to drive at the time of the accident, but who hassince been phobic of driving, despite the fact that she wants to learn. She hasjust discovered that the lawyers representing the defendant have obtained the daughter's therapy records which include the information that the daughter is a victim of incestual abuse, and will question the daughter about this in court. She is so concerned about the possible emotional trauma that this might cause the daughter that she is willing to drop the suit. At the same time, she is extremely angry at the driver of the other car and the insurance company who is defending that person despite repeated violations of driving while intoxicated that she wants retribution for her and her daughter's suffering. She is also still feeling guilt and responsible that she had not prevented the original sexual abuse which occurred when the daughter was a young child, but about which the daughter had only confided in her shortly before the time of the accident. Although these thoughts of being a failure in everything, and the accompanying feelings of guilt, worthlessness, and sorrow had been prominent tin her awareness for some days, it was obvious that she has fought against giving into them. Her defenses against the full experience of these overwhelming negative feelings had allowed her to function adequately on the job and in her daily life. The director, the setting, the protagonist's previous experience with psychodrama, all invite her to relinquish her energy-sapping defenses--to feel the full impact of her feelings of failure and guilt.

She struggles to reveal just how terrible she really does feel about herself. She can't quite seem to express it to her complete satisfaction. A group member volunteers to help her from the role of a double and joins her on the stage. The double takes a position beside her, assumes her posture and motions and begins to echo some of her expressions. It seems to help.

Then the double says, "Sometimes I think I should never have been born!"

The protagonist almost collapses. "Oh, yes!" she blurts out. "That's it! That's exactly what I feel. I should never have been born!" She says it not just as a figure of speech, but as a fact. It is obvious that she means literally and feels deeply that she should never have been born.

The director encourages her to deepen the experience and not fight against feeling her emotions, especially her repeatedly expressed sense of being responsible for every bad thing that has happened to her or to those whom she loves. Then she asks, "When is the first time you can remember feeling responsible for everything?"

The protagonist recalls a scene which occurred when "I was about 3 years old." She is in her bed with a terrible nosebleed and the doctor is paying a visit. Her pillow is covered with blood which is also getting on the sheet. Her mother has just changed the bed and she is feeling very badly because it is getting messed up again. Also, she feels responsible for the nosebleed because her nose had itched and she had rubbed it. She has also lied, denying to the mother that she has touched her nose. The scene is re-enacted psychodramatically and the protagonist's feelings of being bad are clearly portrayed.

The director asks her to reverse roles with the mother and in an interaction between the "mother" and the protagonist, as well as in response to questions from the director, a wealth of information emerges. First, in the role of the mother, the protagonist produces and experiences a terrible anguish verbalized in a fear that she is going to lose the daughter through death. The little girl has been sick almost from the day she was born. "I wasn't supposed to have her, you know," the mother states.

"Tell me about it," the director invites.

The mother had been a member of a religious order, a nun. Unable to accept the discipline, she had left the order and married, in the face of dire predictions from the order that God would punish her severely for such sinfulness. She had her first child, a son, and because of health complications, was warned by her doctor not to have any more children. However, she wanted a daughter, and the same stubbornness which made life in the convent unbearable led her to flout her doctor's advice. She became pregnant with and delivered the baby who was to become the protagonist. The baby had health problems from the beginning. It was believed later that excessive use of x-rays during the pregnancy was responsible for some of the problems. At the time, however, the mother remembered the nuns' warnings of God's revenge.

73

At the time of the nosebleed, the mother was experiencing problems with being close to the child because of her fear that "God will take her away from me." Another theme develops. She had wanted a daughter and she had a daughter. But it is not the daughter she wanted! This daughter is as willful as the mother in terms of having her own way, and absolutely rejects the mother's attempts to make her into a passive, feminine being. She is an unmitigated tomboy even by the age of three. The director takes the protagonist out of the psychodramatic action for a conference. The question is what needs to be done to set things aright. It seems clear that the protagonist must have gotten the message from the mother that she shouldn't have been born. The source and how the mother got the message also seems apparent. The protagonist and the director decide that the protagonist should take the mother for an encounter with God to clear up what appears to the protagonist to be a lot of misinformation about God that the mother got from her religious training and which has had a negative influence upon the relationship between mother and daughter, and still has the power to influence the protagonist in unhealthy ways.

The protagonist has had an encounter with God in a previous psychodrama, and knows just how she wants to set the scene for this encounter. God (for this protagonist) is to be found in his large and rather impressive office, one wall of which is floor to ceiling windows, opening upon a magnificent view of the entire universe. God is rather pleased with his handiwork and expects visitors to be quite impressed, overwhelmed even with the view. They invariably are.

The protagonist explains the reason for the visit. "I am sure that you would never punish someone by taking away their baby ," she explains, "but I don't think my mother knows that. Was my childhood illness a punishment for her for leaving the convent?"

The protagonist is reversed into the role of God. "God" looks at the mother and takes plenty of time in answering. "No matter how far or deep I look, I cannot find any anger at you. Some people don't belong in a religious order.You didn't.I never wanted to punish you for anything, certainly not for leaving the convent. I don't know why you stayed in there so long!"

74

The protagonist reverses roles and becomes the mother. The dialogue continues until all the questions that the protagonist can think of in the mothers role are answered from God's role.

Then there is some more dialogue between God and the protagonist with the director asking "God" (with the protagonist in that role) to address the protagonist's question, "Should I have not been born?"

The answer is a strong affirmation of the protagonist's existence and reassuring. The protagonist should indeed have been born. Next the issue of the decision about continuing with the lawsuit is broached. "God" (played by the protagonist) declines to offer any suggestions in this matter, stating that it isn't his policy to get involved in this level of matters. He does reassure the protagonist that either decision can be a constructive one, that there isn't a clear right and wrong in the issue.

The director again checks with the protagonist, whose energy has increased considerably over the hour or so of the drama. The protagonist is feeling much more confident in herself. She no longer sees herself as having bungled everything she has touched.She feels empowered.She feels like she can make the decision about the lawsuit when she needs to, and is again ready to confront her life with all its issues, positive and negative.

The drama ends and group members share a wide variety of experiences of their own with the protagonist.

This is not an uncommon pattern for a psychodrama. A disturbing situation or event in the protagonist's present life connects affectively with an earlier experience. There is a psychodramatic re-enactment of the earlier experience which clarifies both that event and the current emotional reaction.Some kind of reparative scene is typically enacted next in which corrections are made or an injustice is righted. Often, the protagonist is then returned to the initial scene, and the protagonist notes changes in feelings, and demonstrates new and different ways of dealing with the situation which initially raised difficulties.

Some Issues to Explain:

This psychodrama raises a number of questions which we would like to be able to explain. For example, the protagonist's rather extreme despair about herself calls for explanation. Most of us can extrapolate from our own experiences and understand how the protagonist might be feeling very negatively about herself in the face of so many negative messages from various areas of her life. However, those group members who know something about the life experience of this particular individual are also aware that she has accomplished a great deal in her Iifel and often while struggling against tremendous difficulties. The feeling, expressed in the statement, "I should never have been born!" is obviously a great over-reaction, even though the protagonist feels overwhelmed with the current problems.

A second event to be explained is her strong emotional reaction to putting her feeling into the words: "I should never have been born!" This expression seemed to reverberate as though she had come upon a great and profound truth. It was as though this terrible thought had somehow or another laid dormant in her mind, unspoken but waiting to be recognized. Her reaction was much the same as the common one of reading something important and recognizing that someone else has spoken your "unspoken" thoughts. Someone else has expressed the thoughts and feelings that you recognize as *your* truth. Of course subsequent events in the psychodrama suggest that this may have been a part of the protagonist's truth, unspoken but none the less experienced, since

early childhood. We need then to account for how such thoughts/feelings can lie dormant as well as how they can be reawakened.

We need to ask how the protagonist happened to recall from very early years her memory of the nosebleed scene. Although this protagonist was not specifically asked about it, a lot of experience with this kind of recall suggests that the protagonist had probably not been aware of this memory, or even that it was a memory until asked to remember an early time in her life when this feeling was present.

Other questions have to do with the impact of recalling and re-enacting the early scene. Again, other psychodramatic experience suggests that the reenactment of the nosebleed scene, in itselfl would have a salutary effect on the protagonist. It is also my belief that the reparative scene, the visit to God, intensified and consolidated that effect. The question is: Why and how does such an obviously fictitious, fantastic experience like this produce the change which we saw in the protagonist.

Dimensions of Experience: A Working Guide to Mental Processes

There are numerous ways of characterizing or describing psychotherapeutic methods. There are analytic therapies, supportive and depth therapies, behaviorall cognitive, and affective therapies, "body" therapies, and many others. For examplel more than 10 years ago, over 250 therapeutic approaches were briefly described in a "guide" to psychotherapy (Herink, 1980).

Psychodrama is sometimes referred to as an "experiential" modality of therapy. An experiential modality is typically one in which there is activity other than interview or discussion, the conventional vehicles traditionally as sociated with psychotherapy. The expressive arts therapies (art, music, drama, and dance therapies) are considered experientiall as is Gestalt therapy, bioenergetics, radix and many other methods and techniques in which something otherthan the therapist's verbal interactions with the client is considered to be effective in bringing about therapeutic change. In music therapy, for examplel clients may either listen to music, or perhaps they may make music, singing or playing instruments. In psychodrama, a drama is created in the session. The protagonist, with the assistance of other members of the group, re-enacts events from the protagonist's life, or perhaps enacts scenes that might be anticipated to occur some time in the future, "future projections" in psychodrama language. Although the drama may relate to past, present, or future events, it is the actual creating and experiencing of the drama in the here-and-now that is considered to have a healing effect.

There is, however, another and more important way in which psychodrama can be considered to be experiential. This has to do with the manner in which it deals with the *experience,* the life experience of the client. Now, practitioners of all therapeutic methods are concerned with their clients' experience, of course, although some obviously more so than others. However,therapeutic methods, with few exceptions, look for something beyond, behind, or beneath the patient's experience to *explain* the "psychopathology" or the problems which resulted in the individual consulting the therapist in the first place. Psychoanalysts, for example, as the name implies, set out to analyze and interpret their clients' experience. More specifically, they are especially interested in transference and resistance, unconscious processes which according to theory, if made conscious, give the client insight into his experience and thus, can lead to behavioral changes. Criticizing the psychoanalytic concept and emphasis upon the curative impact of insight, Hobart Mower once defined it as "when the patient agrees with the therapist as to what is wrong withhim. " This sharp comment underscores the fact that the analyst is willing to take certain liberties with the client's experience, indeed to the point of informing him as to what his experience his experience really means.

Many current therapeutic approaches are characterized as behavioral, cognitive, or affective. Behavioral therapists focus heavily upon those actions of the client which are seen as maladaptive behaviors. Usually these are activities which have either gotten the client sent to the therapist, or activities which the client, himself, wishes either to eliminate or to add to his behavioral repertoire. The therapist then tries to establish sets of consequences for the targeted behaviors which will bring about the diminishment of unwanted behaviors, and increase the frequency of desired behaviors, all based on the theory and principles of contingencies of reinforcement. Behavioral therapists tend to eschew even the mention of mental activity, despite the fact that a lot of highly creative mental activity has gone into the development of reinforcement theory and into the identification of the principles of contingencies of reinforcement.

On the other hand, as the name implies, cognitive therapists focus their attention on the clients thoughts. The basic assumption is that the affect and behavior of the client is dependent upon the thoughts that the client has about himself and the world. The aim of the therapist is to test the reality of these thoughts and teach the client more effective ways of coping with negative thoughts which keep the client from living the kind of life he aspires to. Many cognitive therapists take an approach very similar to that of the behavIoral therapists by looking at cognition as behaviors which can then be altered through the appropriate use of reinforcement management techniques.

From the point of view of the affective therapies, emotion is the aspect of experience which needs to be dealt with, and a major goal of the affective therapist is to help the client express unexpressed feelings. The first affective therapy was probably the cathartic technique which Freud learned from Joseph Brewer and which eventually led to the psychoanalytic technique of free association. A tenant of many affective modalities is that the client is suffering from blocked or dammed up emotions, and that getting these feelings unblocked and undammed up will set things right in all areas of clients' lives.

The psychodramatic practitioner, on the other hand, works and thinks in the realm of experience as *experience.* And the psychodramatic stage offers a unique way in which the experience of the protagonist (client) can be presented, explored, and understood.

It is a basic proposition of the psychodramatic method that we are understandable as the product of our accumulative life experience. What we have experienced and how we have experienced it plays the major role in determining how we perceive the world, and how we perceive ourselves in the world. These perceptions, of the world and of ourseIvesl make our behavior in any situation not only understandable but absolutely reasonable, if not inevitable. Thus, if we want to understand another individual we must be able to look at that individual's world and at that individual through that individual's own eyes.

There is a corollary to this axiom. It is not only the behavior of others that is often a puzzle to us. Often enough we do not understand our own behavior. We do not understand why we feel the way we do, or why we are engaging in activities which we know are not good for us. It is concern about understanding ourselves that most often brings us into therapy. In short, it is quite possible for us not to understand our own life experience or perhaps not to be able to make sense out of it. Moreno held that life events often occur too swiftly or too suddenly for us to grasp the full meaning of them, or even to be aware of the full impact of them. We sometimes need a way of exploring our own experience in a more reflective manner, in a way in which we have greater control over what happens or how fast it happens than we do in life, itself. .By allowing us to re-visit, through psychodramatic re-enactment, important moments in our lives, psychodrama provides a method which is uniquely flexible and powerful in the task of achieving greater self-understanding.

Experience

From an objective point of view, we might describe experience as the impact of the ongoing interaction between the individual and his environment. Although we generally tend to think of (to "experience") ourselves as qualitatively different from our physical environment, to draw a rather firm distinction between ourselves and the world that we occupy, to see ourselves as "in" the world, we actually are as much a part of the world as we are in it. Every molecule in us existed in the world before it became a part of us, and will continue to be in the world after we, ourselves, no longer exist. There is a continual exchange go:ing on between the molecules of which we are made up, and the molecules which make up our environment, our world. In every breath we collect millions of molecules of the oxygen with which our world surrounds us, and put back into the world molecules of carbon dioxide in return. Every bite of food, through the digestive process, grabs up some of the world and makes it a part of us, and simultaneously, we are returning to the world parts of it that have been part of us that we no longer need, through defecation, urination, and shedding. This happens continually until we die and then our bodies, which make up a large part of us, become totally part of the environment for those who still live.

The point is that we are *of* our environment, not something that has been placed *in* the world. And there is a continual and ongoing physical interaction between the individual and the environment.

However it is the interaction at a different level which has a much greater impact upon both us and the world, and which is of much more interest to us. This is the interaction between ourselves, individually and collectively, and the world that is mediated through the psychological processes of perception, cognition, emotion, memory and action. This interaction has a lasting impact upon both us and the rest of the world which is not us. We are continually undergoing changes as a result of it. The world is continuously changing, too, as a result of its interaction with us. The change in us influences everything that we do, as well as how we perceive the worldl how we think, what we remember, and how we feel. It is this impact of our interaction with the world that is experience. It involves energy patterns from the environment impacting sensory receptors, in turn setting off neurological activities, some of which we call perception, others labeled cognition and/or memory, still others emotion, and eventually resulting in action on the part of the individual with respect to the environment.

Perception, cognition, emotion, memory , and action, as dimensions of experience, are also continuously ongoing processes. In the field of scientific psychology, these processes are traditionally studied separately. Much effort has gone into the attempt to relate mental activities, such as perception and memory to aspects of the physical world which stimulate them, and into developmental questions such as how the capacity to think develops with the age of the individual. From the most rigorous empirical point of view of scientific psychology , human behavior is considered to be controlled completely by factors in the environment, and the assumption is that if we were to know enough about this process, and could adequately define and measure those factors in any given situation, all human behavior would be controllable through manipulation of the environment.

However, this is not the way we experience life. We do not experi~ncelife as an even, ongoing flow of experience. We are not. aware of perception,cognition, emotion, and memory as the continually evolving rocesses which we know they must be. We do not consider ourselves as simply reactive to our environment. Instead, we experience life as a series of events, experiences, happenings, discrete if sometimes prolonged events, having a beginning and an ending, perhaps overlapping, sometimes flowing into one another. We identify periods in our lives, "when I was a child....," "When I worked at the Mental Health Center...," places and events, "The trip I took to London.:.," relationships, "I met Dr. Moreno in 1960...," etc. Rather than experiencing, we are aware of experiences; rather than perceiving, we are aware of percepttons; rather than thinking, we are aware of thoughts, concepts, images; rather than remembering we are aware of memories; rather than an ongoing, ever changing process of feeling, we are aware of emotions; rather than engaging in behaviors of one type or another, we are aware of actions and accomplishments. We also experience ourselves as evaluating our environment, making decisions, some-times reasoned, sometimes impulsive, and acting upon them as well as upon the environment.

The fact is that it is our own mental activity which creates experiences out of the ongoing process of experiencing. Experiences (and perceptions, thoughts, feelings, memories, and actions)serve to give some kind of stability of a world which is constantly changing.(Cf. Dewey, 1934) They are dynamic processes which have characteristics of being both fixed and fluid. Although we can distinguish between them, for the most part, and can cognitively separate them for study and comment, in life (and in psychodrama) they come thoroughly mixed together, and it is probably more accurate to consider them to be different dimensions of experience rather than different processes.

In psychodrama, one deals with the protagonist's experiences as they are experienced by the protagonist--and this is what makes psychodrama truly an experiential method. It is also what psychodrama shares in common with the philosophical approach, phenomenology, in which great emphasis is put upon phenomena as they are experienced, holding that the key to understanding human existence lies in understanding our experience of the world from this perspective.

And now, having said that perception, cognition, memory, emotion, and action are inseparable dimensions of experience, we are none-the-less going to consider each process individually to emphasize certain aspects of experience. Our discussion will not be the traditional psychological one, relating mental activity to environmental stimuli. Rather, we are interested in the experiences of perception, cognition, emotion, memory and action, and how they interact in the process of experiencing. In short, we are interested in how being who we are means experiencing things the way we experience them, and doing what we do.

Perception

Literally, perception means "taking in." Perceiving is the initial process in our interaction with the world. It has to do with the impact that the world makes upon us, and involves organizing sensory input into percepts, perceptions, and concepts. It is the process that changes the world from the "blooming, buzzing confusion" that William James imagined the infant to experience, into an arrangement of things and relationships with meaning and value. We perceive not only things, but relationships between things. We also perceive ourselves. And we perceive others, which is different than perceiving things because we recognize that others also perceive, and that they are perceiving us just as we are perceiving them. Additionally we perceive relationships between ourselves and things, and we perceive interrelationships between ourselves and others.

Perceiving can be thought of as the basic process of experiencing and perceptions seem to be the basic "material" of experiences. There is some temptation to draw analogies between perception and technology, comparing visual perception, for example, to photographic processes, or aural perception to recording technologies. There are considerable differences, however, and these analogies break down very quickly. The recording of light on film or of sound on tape is a much more passive process, and once recorded, the records are fixed and do not change. This is not true of perception.

While perceiving may appear, at first glance, to be a passive process of simply being aware of what is "out there, outside myself", it is actually a very active process by which energy patterns arising from "what is out there" are organized and assigned both meaning and value. The energy that perceiving can take is apparent to everyone who has tried to perceive (understand) an advanced mathematical concept or an intricate philosophical point. By the time we have become adults, perception of the physical world has become so familiar to us and we are so continuously engaged in it that it is hard to imagine the effort that was originally required to be able to identify the infinite number of things and the relationships between them that surround us. Individuals who have been blind since infants and who have gained sight through surgery learn how much effort goes into integrating vision into their perceptions of simple things like geometrical shapes.

Perceptions are dynamic entities and subject to change and modification over time. This is particularly so of perception of relationship.

Others, initially parents, assist us in this process of learning to perceive.In particular, we are given names and labels to apply to all the variety of things of which our world is made up. We are often coached as to what we should pay attention to, in other words, as to which things we should perceive as well as to what they should mean and how we should value them. In the long run, however, each of us create our own perceptions of both the physical and the social environment. An important part of this process is attributing meaning and value to that which we perceive, and meaning and value always have very subjective aspects.

Despite the considerable universality of human experience, each of us, in a sense, through our perceptions of the world, of ourselves, and of our selvesin the world, creates a unique and individual world.

The "products" of perceiving are variously called percepts, perceptions, or concepts. They are the stuff of thinking and of remembering. They are dynamic entities and are subject to change with continuing experience.

Perceiving is inextricably enmeshed with remembering, thinking, and feeling. Memory plays a part in the formation of perceptions since it is repeated interaction with things that gives rise to perceptions. And, of course, it is perceptions and concepts which we remember when we are talking about memory. Thinking involves reflecting upon perceptions and manipulation of percepts and concepts. Emotions and feelings are aroused by the perceptions to which they are attached, and we know our own feelings by perceiving them.

Probably of greater importance to the individual than his perceptions of the physical environment are his perceptions of the social environment, of the people who surround him, and of his relationships with them. This is because basically, humankind has pretty well mastered the physical environment. That is to say, the technology has been developed which can protect us from most of the life threatening aspects of our physical environment, as has the technology for meeting basic needs such as food. Even though it may not be available to every human being, the technology does exist. It is society that generates most of the serious problems with which we have to cope. Most of the psychological pain that we experience comes out of our relationships and interactions with others. Many of those problems arise from inadequate or inaccurate perceptions of each other.

Two kinds of perception are involved, perceptions of others and perceptions of self. Of the two, the latter, commonly referred to as the self-concept, is probably of the greater interest to the therapist.

There are countless occasions for us to perceive ourselves in relationship to the rest of the world, and these perceptions frequently involve varying degrees of affect. The feeling associated with early self perceptions usually is the result of the reactions of parents or parental surrogates who are either praising us or criticizing us for something. We are told that we are "good" or that we are "bad," and subsequent behavior on the part of these important figures makes it clear that it is better to be good than to be bad. The self-concept develops out of all of these perceptions, and is subject to constant revision in the light of new information. New information might be from the reactions of those others with whom we are currently interacting, or it might also be a revised perception of the criteria for evaluating oneself positive or negative.

Problems involved in the perceptions of others derive from a variety of sources. Accurate or valid perceptions of others means that what they do generally is not terribly surprising to us. The better we understand another, the more their behavior seems to be "in character." In short, valid perception of an other usually means seeing them as they see themselves. Misperception occurs when we see others as we see ourselves. That is, we anticipate that the other will behave as we would if we were in the situation that they are in.

A second source of misperception comes when something about another person, A, reminds us of someone else we have known, B, and we anticipate that A will behave as B does. This is, of course, projection or transference.

A third source of difficulties comes from the fact that most of us attempt, with at least some success, to be less than transparent to most others,and with good reason. We do not live in a world where one can be totally open about oneself without running a grave risk of being exploited. In any event, most of us attempt to keep some of our desires and feelings secret from the world at large, and most of us have some self-perceptions which we are not willing to share freely with others. And most of us have done things which we do not easily tell others about. All of these secrets combined make it more difficult for others to perceive us accurately, and the more of ourselveswe are keeping secret, the more difficult it may be for others to "read" us with a high degree of accuracy .

Finally, there is the fact that all of us are capable of change. When we encounter an individual whom we have not seen for a considerable period of time, we anticipate that person being much the same as the last time we met. We perceive him in the same way. However, that person's experiences may result in his being quite different in any number of ways, hence leading us into incorrect perceptions of him.

Perception is not only the basic process of experience; it is the prototypal creative act. As a matter of fact, that which distinguishes the genius from the rest of humankind is an act of perception. It involves seeing what no one has seen before. Thus, if we accept the story that Sir Isaac Newton was inspired to formulate the law of gravity by the sight of seeing an apple fall down from the tree to the ground, we can say that what he really saw was that the ground fell up to the apple--a little bit. Lots of people had seen apples fall to the ground; nobody had seen that the earth had moved toward the apple.

Cognition

This term covers a lot of activities to which a variety of names have been given. Thinking, reflecting, cogitating, imagining, fantasizing, etc., ar esome of them. A lot of cognition involves an internal dialogue and visualizing ,and is going on continuously in most of us, whether or not we are aware of it. Thinking is often conceived of as symbolic activity which allows us to manipulate things, or the world without going to the trouble that action might entail. So, we can "picture in our minds," an interaction with somebody important to us and "see how it might come out," for example. We can think about our experiences in psychodrama, and consider what role emotion plays in the development of a psychodrama. We can imagine what might have ehappened if we had chosen a difference course of action with respect almost anything we have done.

Imagination and fantasy may be the most fascinating mode of cognition, and certainly have caught the interest of psychotherapists since Freud originated psychoanalysis. Literally, imagination is "the forming images or concepts of what is not present to the senses." Imagination and fantasy allow us to extend ourselves far beyond any limitations imposed by sensory experIence.We can, and do, anticipate the consequences of certain actions we want to take. We solve problems by imagining what would happen if.... We prepare for roles we are to assume by imagining ourselves taking them. We satisfy needs and desires that cannot otherwise be satisfied in the world at large. We express ourselves in our fantasies in ways that the world doesn't allow. And we create through our imagination. There is certainly truth to the notion that we cannot do that which we cannot imagine.

There is not nearly as much difference between imagining and the kind of thinking that is based upon perceptions derived from external sensory input as many people would like to think. It requires a great deal of effort to teach children to distinguish between "fantasy" and "reality," as prescribed by our culture, and one can never be absolutely sure if some thoughts, perceptions, or feelings don't arise from one's fantasies rather than the outside world. Moreno recognized this softness of the boundaries between reality and fantasy in a daring concept, "surplus reality ," a term which refers to fantasy ,imagination, delusions, hallucinations, and other phenomena which lie outside that consensual body of knowledge, perceptions, concepts, and beliefs referredto as "reality," and are considered, therefore, to be unreal. This point of view holds that reality is no more, no less than that which members of society agreeit is. Surplus reality, on the other hand, is a more personal reality, and consists of what the individual "knows" is true, with or without proof, andwith or without the consensual validation of one's fellows. Any mental health worker who has tried to argue a paranoid person out of that individual's delusions will recognize that the delusions are more real to that person then is the mental health worker. On the positive side, surplus reality allows for a fuller expression of the self than does reality, and plays a significant role inthe process of creativity.

The demands that our culture put upon us to adhere to reality are, in actuality, rather stressful, and generate the widely experienced desire to escapeinto an "altered state" of consciousness. This is a state in which the boundaries between thinking and imagination become much more permeable, in which "disbelief is suspended," and there is more integration of the cognitive processes. Altered states, which some believe to be the natural state (e.g.Weil1972,) are induced in many ways. Hypnosis, meditation, music,alcohol and other drugs are just a few of them.

Psychodrama is another. The altered state is labeled the "spontaneity state" by Moreno, and when achieved, results in the suspension of disbelief and the creation of a different reality which is quite intense.

Not only do our perceptions play a major role in cognition since it is our perceptions which make up our thoughts and images; thinking, and especially imagination generate perceptions of their own! These perceptions are just as valid as are those generated through sensory experience. That is, they function in the same way, may stimulate strong emotion, can become memories, influence our actions, and so forth.

Memory

Moreno claimed to have originated the concept of *hic et nunc*, the here and now, and it is almost certain that he introduced it into the field of psychotherapy. Essentially, it says that only the present exists as far as human experience is concerned. The past and the future are essentially fictions. And yet, much of the time in psychodrama, we are re-enacting past experiences. And often enough we are enacting projections into the future. How do we explain this apparent contradiction?

Memory, like other functions, is a process. We recall events in our past, and we think of it as if we are pulling a card out of a card file and reading what we wrote 5, 10 or 25 years ago. It is really not like that. A memory is a current representation of what happened to us before. It is quite plastic, and begins changing almost immediately. For example, by taping psychodramas, we know that the way we remember what happened this morning is different than it was recorded by the mechanical device. How much will it change 5 years from now? Or 10 years? And yet,

I can remember in grea tdetail psychodramas of which I was a part from nearly 30 years ago. Memory is a mental representation in the present of events from the past. We might say that it is our perception of what happened to us in the past. For our purposes, memories are important because they play an integral part of our perception of the present, and because of the interaction between emotion and memory. Experiences in which strong emotion has been aroused tend to be remembered more than experiences in which their has been little emotional arousal. However, some experiences in which emotional arousal has reached extreme levels may not be remembered in the sense of being recall able. These experiences may still have a strong impact upon the ways we perceive the world or ourselves in the world.

Memories of past events, even if they are not accurate representations of the actual historical event, are perceptions of our past experience, and are valid as such. They have the same impact and role in our experience and existence as do perceptions of the present.

And, of course, we can "think over" our memories of past events. Sometimes in doing this, we "see things differently." Thus, memory and cognition may lead to new perceptions, as well as an alteration in the memory.

Emotion

There is always some degree of emotional arousal. Even when we do not think of ourselves as emotionally aroused, there is some feeling tone or mood which can be described by one of the emotional labels or metaphors. People differ in their sensitivity to their own feelings, and some are much more adept than others in identifying this ambient mood or feeling tone. Others, who may be considered "alexithymic," have difficulties in identifying feeling at all, or if it is sufficiently strong, they may have difficulties in differentiating one feeling from another.

The emotions serve as an information processing system. Emotion always has an object; that is, if we are angry, we are angry at something or. someone. If we are afraid, we are afraid of something or someone. There is always a perception to which we react with emotional arousal.

The emotion which is aroused in response to the perception of something represents an evaluation or appraisal of that which is perceived. The appraisal is consistent with the emotion evoked. If that which we perceive is perceived as dangerous, the emotion is fear; if the appraisal is that there is a barrier between us and something we desire, it is anger; if we perceive that something is valuable to us, the emotion is desire, etc. Emotion generates a message to do something about the situation in which it occurs, and this is consistent with the emotion we are feeling also. Thus, fear tells us to get away from the object which evokes it; anger tells us to get something or someone out of our way; desire says, "Possess that," the object of desire.

When we carry out the instruction of the emotion, we are "expressing emotion," and this serves to communicate to others 1) that we are emotionally aroused; and 2) what emotion is aroused in us. In addition to the large movements which express emotion, such as running away or hitting and pushing to express fear or anger, there are facial expressions which communicate our feelings.

The function of emotion, then, is to change our relationship with the world or our environment, at least as we perceive it. Emotional arousal tells us that something has changed and needs tending to, or, on the other hand, that everything is just right so maintain the status quo. It also gives us a suggestion as to what is wrong and what to do about it. The expression of emotion leads to a change, and that change results in a change in our emotional arousal.

That change can be a reduction of arousal. We see a favorite food ,experience a strong desire to have it, follow the instruction to possess it, and the desire abates. The change can be a change in the emotion which is aroused. We find that we have wandered into an area of a strange city which feels unsafe. We experience fear which tells us to get out of there, hail a taxi, and experience relief when the cab stops and picks us up.

And the change can be an increase in emotional arousal. We believe that our house has been unfairly valued by the tax assessor and are angry. We attack by appealing the assessment to the appeals board which, despite the fact that we have proven that our assessment is higher than comparable property , turns us down. We get even more angry.

Emotional behavior is present in the newborn infant in rudimentary form and continues to develop and differentiate throughout one's lifetime. In theinfant, emotional behavior is the first form of communication between the infant and parents. Initially, there is little or no pause between the arousal of affect and its expression. However, this state of affairs does not last for long, and early in life the individual is taught that one must "control" one's feelings ,that one must "think before acting." In short, we learn to inhibit the actions which emotions tell us to carry out.

This training occurs differentially. There are some emotions which w eare taught to suppress more strongly than others. In most families, expression of anger (especially at the parents), fear (especially for little boys), and sorrow are subject to suppression or even punishment.

The emotional process then includes an intellectual or cognitive dimension. Our perception of ourselves in our world arouses emotion, subjective awareness of it, and a tendency to act, which is inhibited momentarily as we intellectually assess the situation, and either follow the original action message of the emotion, or counter it with an different one. The adaptive value of this is obvious and has given rise to the high value we place upon rational behavior, often as opposed to emotional behavior. We may see later, however, that valuing is emotionally based, even if it is rational behavior that is valued.

The point to be made here is that we learn how to block the expression of emotion, and eventually we learn how to avoid the experience of emotion. This involves the development of the intellectual maneuvers which Freud thought of as "defense mechanisms." Originally conceived of as defending us against "instinctual desires," defense mechanisms can defend us against experiencing our own emotions, or against thoughts or memories which are connected with emotion. Defense mechanisms are all too often considered from a negative perspective. "You are being defensive!" is a pejorative accusation, and implies that the target of remark is not being honest. It needs to be emphasized that defense mechanisms have a powerful adaptational function in that they make it possible for the individual to function adequately in the face of strong emotional stimulation. They only become liabilities when they become habitual ways of dealing with emotion arousing situations.

Action

Psychology currently is commonly defined as the "science of human behavior" these days, and certainly a considerable proportion of all the work carried out in this field has to do with behavior. Of central interest, perhaps, is the field called learning theory which has primarily to do with how behavior changes as a result of the interaction between the individual and the environment. This, of course, reflects a universal curiosity about why people do the things they do, and the fact that there is curiosity about it implies that much of behavior is a mystery to us.

As was so with the other processes we have discussed, we will be interested in a particular dimension of behavior, "acts" and "action. Moreno thought in terms of "Man, the actor and interactor," and borrowed the term "role" from drama to help explain everyday behavior. His concept of role emphasizes the degree to which one's behavior is influenced by social factors, while permitting us to account for the infinite variety of individual improvisation. From this perspective, being in a role or acting from a role does not imply that one is dissimulating or being dishonest. Quite to the contrary, Moreno considered personality to be one's role repertoire, and roles to be the means for self-expression.

Perhaps the most important aspect of the role concept is that it implies social interaction. Thus, there can be no role of parent without an other in the role of child; no role of teacher absent the role of student, etc. Because o four interdependence with others, all human behavior can be construed as role behavior. Even the hermit must have others from whom he isolates himself. Roles involve perception, both of oneself and of others, as well as of the relationships between roles. This perception includes ideas of the appropriate actions which one should engage in within a particular role relationship. There are also strong feelings connected with roles. There are roles to which we are strongly drawn, and others which we will go to great lengths to avoid. Role perception begins at an early age, and young children can demonstrate acts and actions which are associated with common social roles.

Action, of course, is motivated by basic physiological drives in part. Emotion, as previously noted, also has a motivating aspect. Even acts motivated by a drive, such as hunger, tend to be modified by learning and emotion. Even though I am hungry, I may go to considerable effort to satisfy the hunger with food that I am especially fond of, and I may even tolerate a high level of hunger if the only food available is particularly distasteful to me. Action, of course, modifies our relationship with our environment, and this can influence all of the processes which we have thus far discussed. Likewise, any of them are likely to be playing a part in determining the action which comes next.

We also perceive ourselves in action, think about acts in which we might engage. Action can stimulate emotional arousal as well as express it, and we remember actions we have taken in the past.

Conscious and Unconscious Processing

With all the activity going on, we are in the position of the spectator at the huge circuses of yesteryear which sometimes had not only three rings, but two platforms between them. There was always too much happening for any individual to pay attention to all of it.

Awareness is like a spotlight that focuses upon one aspect or another of these mental processes. Awareness is associated with cognition and cognitively directed ("willed") action in particular and when things are going along smoothly. Strong emotional arousal intrudes into awareness, demanding that we turn our attention to the object of it as well as the action that is called for by the emotion. And when we have occasion to remember, the memory is"brought into" awareness.

When we are engaged in problem-solving, awareness is like the work-bench in front of us upon which we place the various aspects that we are engaged with.

In any given moment, our awareness tends to be focused primarily upon either that which is going on in our environment, or it is focused internally. Awareness is sure to be focused outwardly when we must engage actively with something or someone in the environment. It focuses inwardly when we are struggling with problems of reconciling contradictory perceptions, or emotions, or other "internal" problems.

Much of our experience is going on beyond the small circle of awareness. That is, it is unconscious. Some of this perceiving, feeling, remembering, thinking, and acting is within easy reach of awareness, and we can become conscious of it simply by being asked to change our focus of attention, or by giving ourselves that suggestion. Sometimes we are aware of unconscious processing when we have occasion to recall it. Sometimes an hour or more of psychodrama is devoted to recapturing all of the processing which was going on in an event of only a few minutes duration.

Although few people find merit anymore in Freud's conceptualization of the Unconscious with its barrier between it and the Conscious, there is little question that events or parts of events can be repressed. That is, they are actively rather than passively forgotten, and the memories associated with them can be brought into awareness only with considerable difficulty.

The Process of a Psychodrama

At this point, it is time to re-examine the psychodrama described above to see if we can answer some of the questions we have asked about it.

As the drama begins, the protagonist is and has been fighting against feeling the full impact of her emotions. She has been defending herself against the impact of negative feelings about herself. That has served to allow her to function in a number of roles in which feeling the full extent of her negative feelings might very well interfere seriously with her performance as a mother, an employee, etc. However, she is now in the role of the protagonist, a role familiar to her. Now is the time when she can acknowledge all her bad feelings, a place in which she can allow herself to feel the full extent and impact of them. However, it is still a struggle to acknowledge the ultimate feeling, and she needs the assistance of an auxiliary. Why?

The answer may be in the statement, "I should never have been born!" which seems to be expressing a number of feelings, guilt, shame, despair, pain, loneliness, inferiority and possibly others. If we ask ourselves what action message this conglomeration of feelings is giving her, the answer may very well be: "Don't exist!" It may very well represent some preliminary motivation toward suicide. If this is so, her resistance to acknowledging it is a positive indication. It implies that there is also a strong to live counteracting any thoughts that she shouldn't be alive. Assuming that this struggle has actually been going on, it has been beyond the focus of awareness.

Her response to the auxiliary's production of the statement, "Maybe I should never have been born!" strongly suggests that the auxiliary has correctly perceived her state of mind. Contrary to the fears that some professionals express that the protagonist may be overly influenced by suggestion when a double is utilized, with very few exceptions protagonists immediately recognize when the contribution of the double does not fit with their here-and-now thinking and feeling. Protagonists seldom have problems in rejecting such expressions, and when the double's offering is close up not exact, the protagonist is usually able to re-express the statement to fit more accurately. The fact that another individual can act as a double tends to affirm the notion that emotion can communicate non-verbally.

Once the protagonist expresses the depth of the feeling, the last remnant of defense is relinquished, and the subjective feelings of her emotional process reach their fullest. She nearly collapses and the director might have facilitated an even stronger non-verbal expression of the feeling.

Now that the emotion is fully represented in her awareness, the director asks the protagonist to identify a source of this feeling. The common questions that are asked are: "Is this a new feeling or have you ever felt this way before?" "When have you experienced feeling this way before?" "When was the first time that you can recall having this feeling?" etc. Time after time, protagonists can identify the emotion as a familiar one and recall other situations in which it occurred.

Memory recall is heavily influenced by affect, and affect is often the bridge between current experience and the past. One function of affect seems to be to remind us of previous experiences, and this may well where the action messages of emotions come from. It is as if affect is saying, "Hey! You've been here before. This is what happened and this is what you need to do about it." People generally report that memories from the past which standout are those which are associated with a significant degree of affect. Practitioners of Adlerian therapy have placed great emphasis upon the earliest memories that one can recall. They, too, find that the earliest memory recalled in a given situation is one which tends to mirror the subject's current feelings about himself.

One can raise the question of the validity of memory, and especially of these very early memories. The answer is that we are less interested in the history, the actual past experience of the protagonist than we are in the present day perception of the past. It is the past as the individual remembers it that influences behavior in the here and now. This is a phenomenological approach to memory, and with respect to how the individual is living his life in the present, valid in and of itself.

At the same time, enough people have researched their pasts as a result of memories recalled in psychodrama, to give us some confidence that the memories recalled have a significant degree of historical accuracy. And some-times, the psychodrama helps clarify memories. For example in one psychodrama, the protagonist remembered a time when her father had choked and killed a cow. In re-enacting the scene, when the protagonist thought she was perhaps two and a half years old, she saw her father with a cow in labor. She saw him pushing the cows head into a bucket of water and thought that the was drowning it. As she explored the scene, she came to the conclusion that he was merely trying to give the cow a drink of water. The affect associated with the scene included the fact that she was not supposed to be in the barnyard where she watched the scene thru a door, plus her father's anxiety and concern about the cow which seemed to her to be anger at the time. Combined, these made a heavy emotional impression on the little girl, and because of her disobedience, she was forestalled from asking any questions about what she had seen. Thus, she had no way of correcting her perceptions at the time the event occurred.

The next part of the drama involves a re-enactment of the scene remembered. As are most very early memories, this memory is like a snapshot, and has little extension in time. She only remembers being in bed, the mother and the doctor talking outside her door, and the powerful negative feelings about herself for causing her mother nothing but trouble.

The psychodrama permits an elaboration. The memory is only a snap-shot, a slice through time, or, more importantly, a slice through experience. But if we look at a snapshot, we know intuitively that something connected with it came before and that something followed. In psychodramatic reality, we can take the snapshot and re-create what could have led up to this scene and continue it, if that is desirable. We can even, as this director does, re-create events that might have happened but which we know didn't occur at that time. We know and accept that this is not (necessarily) a valid re-creation of history, although it often is experienced that way by the protagonist, but a valid expression of memory and emotion, both in the here-and-now, and connected experientially with the past.

The director asks the protagonist, with the help of auxiliaries, to enact something which did not happen, a dialogue between the mother and the child. Although this didn't happen when the protagonist was three years old ,it is happening now. Moreno called this "surplus reality ," indicating that it is phenomenologically real for the protagonist even though it is recognized to have never happened existentially. The protagonist is asked to portray the mother, to put herself as completely as possible into the mother's role, and through interview and soliloquy, to explore the relationship between her mother and herself from the perspective of the mother. Again, historical accuracy is secondary. Of greater importance is the protagonist's perception of the mother and of the mother's feelings toward her. Very often, we have perceptions of significant others and of their feelings toward us, even fantasies of their feelings toward us, even though these have never come into our focus of awareness. There are lots of thoughts and feelings in most people which remain unconscious in this sense. They can have an influence upon us none the less.

When the psychodrama involves the protagonist as an adult re-enacting a childhood scene, we know and expect that the protagonist will utilize knowledge of the significant other that they would not have or would not be able to articulate as a child. In the present situation, the protagonist has dealt with a number of problematic feelings toward the mother that she carried into adulthood. She and the mother have gained a better understanding of each other and a better relationship in recent years. She is able to utilize this increased awareness of the mother's circumstances in producing this "surplus reality ."

A number of important things are going on all at once during this part of the psychodrama. The protagonist is clarifying her perceptions of herself as a child, perceptions, some of which carry with them a heavily affect-laden thought, "I should never have been born!" this affect carries with it an instruction on the order of "I need to make up for having been born."Because she has defended herself against these perceptions of herself and her relationship to the world, they have not consistently perturbed her. They have been "unconscious" most of the time. But they are still active and can be accessed when the conditions are right. The affect becomes quite influential when things are not going right, leading the protagonist to blame herself for all of the troubles that she experiences.

Simultaneously she is clarifying her perception of the mother. By taking the role of the mother who is carrying a similar negative perception of herself as an individual who has done wrong, who is doing wrong, and who has disobeyed God and her doctor to have the daughter she desperately desired, the protagonist explores the existential position of the mother. She spontaneously experiences a number of nearly overwhelming affects. She experiences a tremendous love for the daughter in spite of the fact that the daughter is headstrong and disobedient. And she experiences the heart wrenching terror and fear that the daughter will die, a fear associated with the guilt that she will be responsible because of her disobedience to God's will. She experiences, as the mother, attempts to suppress her love for the daughter, bracing against the anticipated loss of her, and thinking, "If I don't love her too much, maybe God will not punish me by taking her away."

The protagonist is also clarifying the relationship between herself as a child and the mother. She has always known that she was not the daughter that the mother wanted; i.e., she did not match the mother's fantasy of what she would be like. Additionally, she had experienced a sense of distance on the part of the mother which she had interpreted as rejection. And she had felt that she was bad because she was so disobedient to the mother and that the mother was rejecting her because she was bad and not the daughter that the mother wanted.

In the psychodramatic scene, it becomes clear to her that there is no question of the mother's love for her, and even though the mother feels that she was wrong in having her, perhaps that "she should never have been born," that this is based on the mother's guilt, and is not a rejection of her by the mother because she is bad and different than the daughter the mother wanted. The mother may be sorry that she disobeyed God and her doctor, and feel badly for having gone ahead and had the daughter, but she is not sorry that her daughter exists nor does she wish that the daughter had not been born. The protagonist understands the mother's distance not as rejection but as a defense against the fear of losing her. And she sees herself not as just a disobedient child who should never have been born, as one who only causes trouble for everybody, but as a longed for child, born into a morass of troubled emotions which existed long before her existence and over which she has absolutely no control.

Correcting Old Perceptions

Even though a perception may be gotten out of awareness, the right combination of events can re-activate it, or, more commonly, reactivate the affect associated with it. In that situation, then, the individual may experience strong feelings without being able to account for them.

We could describe what is happening in the psychodrama in the scene with the mother is that the protagonist is correcting old and well defended perceptions of herself, her mother and the relationship between them. First it was necessary to access these perceptions, and this calls for undoing the defensive work that has kept them hidden from view for years and years. Next, the affect must be experienced and expressed. When a strong emotion is experienced, and the action demanded by the emotion is not taken--that is to say, when an emotion has been aroused and not expressed--the individual is left with an action uncompleted. When the affect is re-accessed, the individual has a strong urge to complete the expression of the affect.

When the instruction of the emotion has been carried out, which is often what we mean when we say that the emotion has been expressed, the affect changes. Another one takes it place. Also, when one carries out the instruction of the emotion, one's perception of the situation changes because the instruction of the emotion has to do with changing the situation. Likewise, if one's perception of the situation changes, before the emotion has been ex-pressed, the emotion can also change without a need for it to be expressed. It no longer needs expression. There may be a new action message.

As a child, the protagonist perceived herself on a number of occasions as "bad;" that is, of little worth. This feeling motivated her to act responsibly, to make up for her badness. She also defended against feeling bad. Some of her negative self-evaluation came from honest disapproval from the mother. Another part of it came from the mother's own self-disapproval associated with having had the daughter in defiance of God and her doctor. While she was able to transform part of the negative self-feelings through being concerned about the welfare of others, the part that came from the mother's emotional problems remained untouched by these actions. This is the part that was isolated and pushed out of awareness defensively.

In other words, she had self-contradictory self-perceptions. Actually we all have lots of self perceptions. She had some that were positive and others that were negative. When the situation she was in informed her that she was doing the right thing, behaving properly, etc., which has a lot to do with how people responded to her, the positive self perceptions held sway, and she felt good about herself, was able to function well, etc. When, however, the situation she was in, or the people she related with gave her messages that she was not treating them as she should, doing the proper thing, etc, the negative perceptions with all their negative affect, threaten to emerge, stimulating both the actions associated with them ("Be more responsible." "Treat the person better." etc.) and defensive maneuvers to deactivate the negative perception/affect.

A series of negative messages had occurred for the protagonist recently, and the negative affect had been strongly provoked. She was struggling to defend herself against it. The psychodrama setting provided a situation in which she could relinquish her defensive struggles against her negative feelings. They could be experienced fully.

The emotion connected to the sense of worthlessness and non-entitlement was primarily sorrow. The messages of sorrow are: "You have lost something valuable," (in this case, the loss is of self-worth); and "Give it up." There is no wonder that the protagonist fights against this message. To give up one's self-worth may well be life-threatening. This is certainly implied in the phrase,"Maybe I should never have been born."

So far, only the affect has been pushing into the surface of awareness, and the protagonist has been trying to push it back beneath the surface. In the psychodrama, however, she not only lets it emerge; with the encouragement and assistance of the director, she even tries to pull it further into awareness. It comes, and with it emerges a memory. Even though the memory is no more than a snapshot of the past, it provides enough of a stimulus for the protagonist to recreate a rich portrayal of her situation as a very young child. Through the psychodramatic reenactment, she sorts out the messages she was receiving from the mother, and the source of them. Her perception of the mother, as associated with her perception of herself at that age, is expanded. Experiencing the love of the mother for the child, and the desire for the daughter in the mother's role leaves her with no doubt of the reality of both of these feelings. Heretofore, her previous experience of the mother, when she was age 3, had left her in some doubt about this. Now, using new information as well as psychodramatic experience, and having accessed the perception which has been fended off for so long, she is changing that perception of herself. Phenomenologically, she is correcting or updating her history. Or perhaps we could say that she is correcting her memory of her history.
Either way, the results of doing this should be two-fold. She will need to invest less energy in defensive processes; and situations in which things are not going well for her should be less able to evoke the feelings of worthlessness.

Now let's see if we can understand what part the scene with God played in the psychodrama and what effect it had upon the protagonist. The stimulus for this scene came from the scene with the mother and the

3 year old child. After the protagonist had sorted out the situation, and recognized the mother loved and wanted her, but was afraid that she was going to die as God's punishment of the mother for leaving the convent, etc., she expressed some anger toward the mother. "Why didn't you get this all figured out before you had me? Then you wouldn't have laid this trip on me!" she confronted her mother. The invitation to take mother to get things straightened out with God can satisfy this anger in a positive way. Besides, the basic theme of the drama, "Maybe I should never have been born!" is a transcendent issue. Who has the answer to the question of whether or not I should have been born?

This part of the drama accomplished several things. It did serve to reduce the anger toward the mother. More than that, it gave the protagonist an experience with the noumenal, the sense of being a part of all that is. The emotion is serenity, and it is a very powerful experience. At the end of the drama, the protagonist was no longer experiencing the negative feelings nor overwhelmed by the sense that whatever she did was going to be wrong. She felt, instead, calm, self-assured, appropriately confident that she could handle the problems that she was facing, and determined not to be taken advantage of by others.

The question that is invariably asked, especially by mental health professionals when they experience the impact of psychodrama upon the protagonist for the first time, "How long will this change last?" The answer to that question is very complicated.

First, affect is always subject to change and is always changing as our relationship with the world is constantly changing, and our perceptions are constantly changing. It is quite possible that this protagonist will, at some time in the future, re-experience significant feelings of self-worthlessness. These might be associated with the time in her life that has been dealt with, or with other life experiences. We have not eliminated the experience; instead we have added a new experience which we anticipate will be called forth if the memory of the original experience is accessed. We anticipate that the new, psychodramatic experience can modify and ameliorate the effects which the original experience has been having upon the protagonist recently. At the same time, we don't know how many other negative self-perceptions may be present which may not have been corrected by this experience. If there are some, and if they were not accessed during this drama, then other corrective experiences may be indicated when they intrude into the protagonist's awareness and interfere with her functioning in the future.

At the same time, the psychodramatic experience, though obviously different from "real life," is, at the same time real life, also. The experiential impact of psychodrama, therefore, cannot be dismissed as less important than any other experience. As a matter of fact, a psychodramatic experience may actually be the most important life experience that someone will undergo.

References

Dewey, J. (1934) *Art as experience*, New York: G. P. Putnam's Sons

Breuer, J. & Freud, S. (1957) *Studies in hysteria*. New York: Basic Books

Herink, Richie. (1980) *The a to z guide to more than 250 different therapies in use today*, New York: New American Library

Moreno, J. L. (1975) Mental catharsis and the psychodrama, *Group Psychotherapy, Psychodrama & Sociometry* 28, 5-32.

Moreno, J. L. (1953) *Who shall survive? Foundations of sociometry, group psycho-therapy and psychodrama* (2nd ed.) Beacon, N. Y .: Beacon House

Moreno, J. L. (1940) Mental catharsis and the psychodrama, *sociometry*, Vol iii, pp 209-244. Reprinted in Greenberg, I. (1974) *Psychodrama: Theory and Therapy*, New York: Behavioral Publications.

Weil, A. (1972) *The Natural Mind*, Boston: Houghton Mifflin Company.

THE CANON OF SPONTANEITY–CREATIVITY

A science of man should start with a science of the universe. A central model of the universe hovers continuously in our minds, if not consciously then unconsciously, whether magical, theological, or scientific. It influences the form the central model of man takes. An incomplete or deficient model of the universe is better than none. (Moreno,1955a, p 359, italics in the original.)

Creativity has long been a philosophical and scientific question. Not only do we stand in awe of human creativity, manifested in the great works of art in all its forms, painting, sculpture, theater, poetry and fiction, as well as in the profound theories of the great philosophers since the days of Ancient Greece, the explanations of the workings of the natural world evolved by scientists, plus the near infinite products of technology by which we have gained so much control over our natural environment, but we also wonder about creativity in the physical and biological realms. How did the universe evolve and how is it continuing to evolve? How has our planet changed from a collection of gases to the foundation of our existence? How did life emerge and how has it evolved? How have new species of all life forms been created? And perhaps, how have we become who we are? As Moreno put it, "Creativity is *the* problem of the universe; it is, therefore, the problem of all existence, the problem of every religion, science, the problem of psychology, sociometry and human relations. indeed, it is *the* problem of the universe (1955c, p 382.)"

J. L. Moreno originated the notion of spontaneity–creativity. "I formulated this twin concept as the primary principles of existence in my earliest effort [*Das Testament des Vaters*] to comprehend the living universe in its entirety (1955b, p 361.)" In this final issue of *Sociometry* that Moreno edited, from which the above is quoted, Moreno elaborated on the theory of spontaneity–creativity which he had first presented in the second edition of *Who Shall Survive?* It was intended in part as a reply to a critique of his earlier presentations of spontaneity–creativity by Pitirim Sorokin (1949), a friend and fellow philosopher who was also a student of creativity. He apparently sent a copy of the article to Sorokin prior to publication for Sorokin provided another critique (1955) which immediately followed the Moreno article.

Here, as previously, Sorokin objected to the linking of spontaneity with creativity. "I am not sure that the marriage of spontaneity–creativity to one another, imposed by Moreno, is a happy marriage that promises a most fruitful study of the properties of each of the married partners and of their progeny (p 374-5.)" He insisted that creativity often occurred without the agency of spontaneity, but the examples he gives reveal his lack of understanding of the Morenean concept of spontaneity. In a reply to Sorokin in the same issue, which we shall soon visit, Moreno attempts to broaden Sorokin's understanding of spontaneity and the warming up process.

Sorokin advocated for the traditional approach to the study of creativity through the analysis of the lives and products of our geniuses. Moreno, recognizing that everyone is engaged to some degree, no matter how slight, of creativity in everyday life chose to examine creativity in the ordinary person and in everyday life:

> Creativity is not a separate mystic, aristocratic, aesthetic or theological category; if it is on top, it is also on bottom; it is everywhere; if it is in the macroscosmos, it is also in the microcosmos; if it is in the largest, it is also in the smallest; in the physical and social atom, it permeates the common and the trivial; it is in the eternal and the most transitory forms of existence; it operates in the here-and-now, in this pencil and in this paper, as I am writing these words to the reader. (1955c, p 382)

The theater of spontaneity became the laboratory for the study of spontaneity–creativity. As Moreno sometimes explained to students of the Moreno Institute, one cannot see the creativity of the painter, the sculpturer, the composer or the playwright, nor even the creativity of the director and actors of conventional theater. Their creating is done in private, in rehearsals and preparation for production. In the theater of spontaneity, on the other hand, there is no backstage, no script, no rehearsal. The spontaneous drama, its characters and their interactions, all is created and produced at the same time, in the moment, in the here-and-now. Working with the spontaneity players, making interaction diagrams, watching how different actors interacted with each other, Moreno began to identify the principles of spontaneity and spontaneity training.[4]

Moreno's notion of spontaneity–creativity is the foundation stone of all of his major theories and methods. To him, it was an insight into the very nature of the universe and it provided a basis for his conviction "that human beings do not behave like automatons but are endowed in various degrees of initiative and autonomy." (1955b, p 361.) In other words, our behavior is not so totally, absolutely determined as the doctrine of natural science would have it.

Because spontaneity was an evolving concept, Moreno's discussions are sometimes inconsistent, a factor which opened him to considerable criticism (i.e., Sorokin 1949; Auculino, 1954.) Moreno pled guilty but asked for leniency, explaining that "it is difficult to convey the full meaning of undeveloped concepts like spontaneity and creativity which are in transition from a pre-scientific to a scientific formulation. "One must use," he says, all available objective and subjective resources." (1965b, p 361-2.)

The Doctrine of Spontaneity–Creativity

In the second edition of *Who Shall Survive?*, Moreno presents a graphic depiction of his theory of creativity, calling it the Canon of Creativity[5]. He reprised this in the December, 1955 issue of *Sociometry*, the last issue that he published before he giving that journal to the American Sociological

---

[4]As noted in the chapter on the history of psychodrama, he also figured out how to use the principles of spontaneity theater in a therapeutic fashion.

[5]"Canon" here is used in the sense of fundamental principles or axioms to distinguish this from a theoretical statement to be confirmed through experiment or observation.

Society. In this article, he elaborated upon his earlier discussion. Moreno's depiction of The Canon of Creativity is reproduced on the next page. The entire universe is infinite creativity, Moreno says, and spontaneity is the catalyst which releases creative processes. On the diagram, he places creativity (C) at the center. A diameter is drawn through the center of the circle. Spontaneity (S) on the left is considered necessary to act as a catalyst giving rise to a conserve (CC,) product of a creative act or event.

Spontaneity can proceed in two directions: one is along the diameter towards interaction with creativity. The other direction is around the perimeter of the circle toward the conserves, and, from the conserves back toward spontaneity. This represents their continual and continuing interaction.

Warming up (W) is defined as the operational expression of spontaneity. It emerges in response to the situation, and may be activated by existing cultural conserves, such as a piece of music, oratory, or other man-made entity. Spontaneity also serves to reactivate and revitalize established conserves.

Now let us take a detailed look at these concepts: spontaneity, creativity, the conserve and cultural conserve, and warming up.

Creativity

Usually we think of creativity as a faculty or talent that one possesses and which is manifested in the creative behavior of artists, writers, scientists and other individuals who bring forth something new, original and outstanding. From Moreno's point of view, however, what we commonly think of as creativity is actually the spontaneous–creative process. In the Morenean approach creativity itself is only one part of the process. "The universe is infinite creativity," he said (1956, p 103.) "Creativity is the ever nourishing maternal center," he wrote, and "creativity is the arch substance....[spontaneity] is the arch catalyzer (p 105, 1956)."

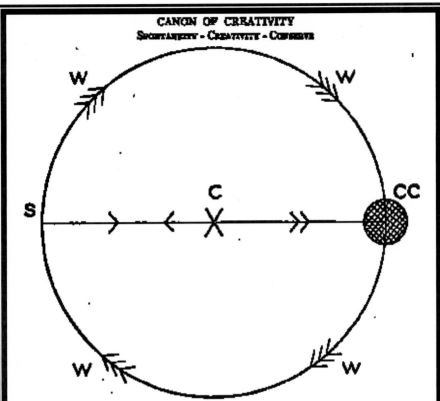

CANON OF CREATIVITY
Spontaneity - Creativity - Conserve

FIELD OF ROTATING OPERATIONS BETWEEN SPONTANEITY-CREATIVITY-CULTURAL CONSERVE (S-C-CC)

S—Spontaneity, C—Creativity, CC—Cultural (or any) Conserve (for instance, a biological conserve, i.e., an animal organism, or a cultural conserve, i.e., a book, a motion picture, or a robot, i.e., a calculating machine); W—Warming up is the "operational" expression of spontaneity. The circle represents the field of operations between S, C and CC.

Operation I: Spontaneity arouses Creativity, C. S——> C.

Operation II: Creativity is receptive to Spontaneity. S <——C.

Operation III: From their interaction Cultural Conserves, CC, result.
  S——> C——>> CC.

Operation IV: Conserves (CC) would accumulate indefinitely and remain "in cold storage". They need to be reborn, the catalyser Spontaneity revitalizes them. CC——>>> S——>>> CC.

S does not operate in a vacuum, it moves either towards Creativity or towards Conserves.

Total Operation

Spontaneity-creativity-warming up act $<^{actor}_{conserve}$

From Moreno, J. L. "Who Shall Survive?", 2nd Edition, 1953, page 46.

Within the context of spontaneity–creativity theory, creativity can best be thought of as potential, both manifested and possible. It includes that which has been created and that which can be created. Whatever is possible belongs to the category of creativity.

But this is not all. If we are willing to relinquish the doctrine of absolute determinism, as Moreno did, it is possible that there are things which might have been created but were not: life forms which might have evolved but didn't, languages and tools which might have been invented but weren't, books which could have been written, music which might have been composed, and so forth, things which *could* have been created–but for some reason (Moreno would say "for a lack of spontaneity") were not. Creativity must comprise all of these things also. It must be obvious to every thinking person that creativity in the world has not come to a halt. Those of us who have been around for a while can remember the emergence of television, jet engines, nuclear bombs and nuclear generating plants, and, of course, the computer and all its accouterments, peripherals, and software programs. If creativity is the archsubstance, then it must also comprise all those things which will be created in the future, new life forms, new products of human creativity, new societies and social orders, new languages, tools, works of art, scientific theories, things of very category.

And finally, assuming again that determinism is partial and not absolute, then the potential for being created is not inevitable. In this case creativity subsumes all those things which might be created in the future–but which will not be. "The universe is infinite creativity."

Spontaneity

Defining Spontaneity

Because his writings do show inconsistencies in his various discussions of spontaneity, we will take primarily a developmental and chronological approach in this section. When the word spontaneity first shows up in Moreno's writings, it is in the context of spontaneity training. For example, in *Das Stegreiftheater*, published in 1923, he proclaimed that spontaneity training will be the main subject in schools in future. At this point Moreno considers spontaneity to be a mental function, comparable to, perhaps, but distinctly different from intelligence. He relates spontaneity to action when confronted with the unexpected. He also considers it to be a function which is susceptible to training:

"The sense for spontaneity, as a cerebral function, shows a more rudimentary development than any other important, fundamental function of the central nervous system. This may explain the astonishing inferiority of men when confronted with surprise tactics. The study of surprise tactics in the laboratory shows the flexibility or the rigidity of individuals when faced with unexpected incidents. Taken by surprise, people act frightened or stunned. They produce false response or not at all. It seems that there is nothing for which human beings are more ill-prepared and the human brain more ill-equipped than surprise....When compared with may other mental functions such as intelligence and memory, the sense for spontaneity is seen to be far less developed. This may perhaps be so because in the civilization of conserves which we have developed, spontaneity is far less used and trained than, for instance, intelligence and memory." (translated and revised from *Das Stegreiftheater* 1923 and published as The Philosophy of the Moment, *Sociometry*, Volume 4, Number2, 1941. Also included in Spontaneity theory of Child Development, Sociometry, vol 7, 1944, and in Psychodrama, Volume I, 1946, p 47.)

Moreno is still regarding spontaneity as a mental function or perhaps a trait when he writes in the first edition of *Who Shall Survive?* (1934) about spontaneity training as a means of increasing "spontaneability" of the residents of the Hudson Training School for Girls. Using methods and techniques which would probably be called role training today, Moreno and the staff had their charges play out numerous scenarios with which they might be confronted when they re-entered society at large. The aim was to develop a broader, widely varied repertoire of actions and greater flexibility in meeting all kinds of situations.

By 1944, when Moreno published the Spontaneity Theory of Child Development (Moreno, J. & Moreno, F.) the notion of spontaneity as an energy begins to emerge. In this paper, the Morenos emphasize the radical change in situation which birth brings about, the infant having moved from a safe, closed, protective, confining, equilibrium to a world with

113

open, unlimited space, from darkness into a kaleidoscope of shapes and colors, light and darkness:

> He moves into this world with such a suddenness, that his successful adjustment is one of life's great riddles. Within a few minutes, he practically changes from one world to another.

Energy, perhaps, but Moreno makes it clear that this is not energy in the sense of biological energy, energy as known to natural science. It is a different kind of energy, an energy which does not obey the law of conservation of energy.

It is also in this article that this formulation appears: "We have called this response of an individual to a new situation–and the new response to an old situation–*spontaneity*." This will soon morph into a definition of spontaneity.

This article, Spontaneity Theory of Child Development, is reprinted in *Psychodrama Volume I* (1946) and is followed by several pages of discussion entitled General Spontaneity Theory. It is a passage which, with minor variations appears in other places, and one which I have never found particularly easy to understand. He asks two questions (p 85.) The first is: "Does the s factor [spontaneity] emerge only in the human group or can the s hypothesis be extended within certain limits to sub-human groups and to the lower animals and plants?" This question does not really get answered. The second question is the primary subject of discussion: "How can the existence of the s factor be reconciled with the idea of a mechanical law abiding universe, as, for instance, with the law of the conservation of energy?"

Moreno uses the psychoanalytic concept of libido as an example of applying the law of conservation of energy to the psychological realm. Freud, he reminds us, assumed that if the flow of libido was deflected from its natural aim, sexual intercourse, that the dammed up energy must flow elsewhere, finding new outlets such as aggression, substitution, projection, and sublimation. Psychological determinism in this formulation is complete. What can't be explained in terms of the conscious can be assigned to the unconscious, and there is no place for spontaneity in such a system. Because spontaneity and creativity can result in levels of organized expression not fully traceable to preceding determinants, Moreno recommends "abandonment or reformulation of all current psychological theories, openly or tacitly based upon psychoanalytic doctrine....(p 87.)"

He answers the second question on reconciling spontaneity with conservation of energy in this way: "It is in the interaction between spontaneity–creativity and the cultural conserve that the existence of the s factor can be somewhat reconciled with the idea of a law-abiding universe, as for instance with the law of conservation of energy." In other words it is in the world of conserves that the conservation of energy and the other laws of physics hold firm.

Finally, in the second edition of *Who Shall Survive?*, published in 1953, Moreno offers another discussion which appears largely to be a re-write of the one just mentioned. It serves as an introduction to the Canon of Creativity. Here he is ready to characterize spontaneity as a "kind of energy." But he seems a bit uneasy with the idea, probably realizing what a bold step it is to hypothesize a form of energy different from energy as known by physical science and as yet unrecognized by scientists. He is quite conscious of the law of conservation of energy and its importance in natural science and how spontaneity seems to refute it, at least in the human realm:

> The universe is infinite creativity. But what is spontaneity? Is it a kind of energy? If it is energy it is unconservable, if the meaning of spontaneity should be kept consistent. We must, therefore, differentiate between two varieties of energy, conservable and unconservable energy. There is an energy which is conservable in the form of "cultural" conserves, which can be saved up, which can be spent at will in selected parts and used at different points in time; it is like a robot at the disposal of its owner. There is another form of energy which emerges and which is spent in a moment, which must emerge to be spent and which must be spent to make place for emergence, like the life of some animals which are born and die in the love-act.
>
> It is a truism to say that the universe cannot exist without physical and mental energy which can be preserved. But it is more important to realize that without the other kind of energy, the unconservable one–or spontaneity–the

creativity of the universe could not start and could not run, it would come to a standstill. (p 47.)

Here we also find the statement: "spontaneity propels a variable degree of satisfactory response which an individual manifests in a situation of variable degree of novelty (p 42.)" A chapter of *The Sociometry Reader* (Moreno et al, eds., 1960) entitled "Spontaneity–Creativity–Cultural Conserve," authored by Moreno begins with this sentence: "*Spontaneity* is the variable degree of adequate response to a situation of a variable degree of novelty (p 8,)" and this would become the standard response whenever Moreno felt called upon to define spontaneity.

Forms and Types of Spontaneity

In the 1946 presentation of spontaneity theory Moreno delineates three different forms of spontaneity. He calls the first dramatic quality. "It is the quality which gives newness and vivacity to feelings, actions, and verbal utterances which are nothing but repetitions of what an individual has experienced a thousand times before—that is, they do not contain anything new, original, or creative (p89.)" Later, after this thinking has become more systematized, Moreno will call this the kind of spontaneity which revitalizes conserves. It is probably closely related to many of the examples of "flow," as spelled out by Csikszentmihalyi (1990.) It involves reproducing learned acts or actions in a fresh, new way. Moreno gives as an example the repetition of a prayer which has been spoken hundreds of times. One may simply go through the motions, so to speak, ritually repeating the prayer. Another may say the words with intensity of feeling so that he speaks the prayer in a way the differentiates it from his own previous repetitions of it. That involves dramatic form of spontaneity.

The next form is called creativity. "One may be totally productive and creative although perhaps undramatic and insignificant as an individual if we would empty his mind we would find it in a permanent *status nascendi*, full of creative seeds always in the temper of breaking up existing conserves and germinating new forms, new ideas, and new inventions (p 91.)" This individual lives in a world of novel experiences and "the s function is not satisfied expressing ony the self; it is eager to create the

116

self." The individual who displays this form of spontaneity makes the most of his resources, his intelligence, memory and skills. He surpasses those who may be superior in resources but who never seem to make the most of them.

The third form of spontaneity is labeled originality. This form is displayed by the individual who freely expresses himself and uniquely but without the novelty of response which warrants the label of creativity. "This is often illustrated in the spontaneous drawings of children and the poetry of adolescents who add something to the original form without changing its essence (p 92.)"

Then, in the discussion in Sociometry, vol XVIII, he offers another classification. Here he suggests three types (rather than forms) of spontaneity and they are:

1. Whenever a novel response occurs without adequacy, that is undisciplined or pathological spontaneity;
2. Whenever an adequate response occurs without significant characteristics of novelty and creativity; and
3. Whenever an adequate response occurs with characteristics of novelty and creativity.

Here again we can see why it is difficult to follow Moreno in his development of concepts, and why he was criticized for his writing style. Kipper (1967) offered some suggestions which he thought might help resolve some of the apparent contradictions and incompatibilities associated with the concept of spontaneity. He found it deplorable that spontaneity "has been often misconceived and misinterpreted" (p 62) in view of its importance in the Morenean system. He argues that the fact that spontaneity has colloquial meaning as well as technical meaning has caused some of the confusion. A second source of misunderstanding comes from failing to differentiate between the theoretical aspects of spontaneity and the practical ones. Finally he suggests that Moreno has assigned too many characteristics to spontaneity, some of which are really features of the warming up process. Elucidating this, Kipper believes, would increase clarity with respect to the concept of spontaneity.

Kipper finds and discusses nine characteristics of spontaneity in five of Moreno's discussions of the concept. The characteristics are listed below with Morenean quotes which support them. Some of the quotes are pointed to by Kipper; others are mine:

Energy: Kipper maintains, correctly I believe, that as energy, spontaneity is not itself directly observable, only through the agency of a carrier. He points out the necessity of keeping the two differentiated.

Spontaneity is unconservable and non-accumulative: Kipper here describes the relationship of spontaneity and the moment, a concept we have not yet mentioned but which is very much a part of Morenean theory and philosophy. Spontaneity exists only in the present, the here-and-now.

Cosmodynamics: Kipper notes that human relationships, from the moment of birth require at least two persons, and that the major emphasis is on dealing with the immediate environment. Later concern develops about cosmic phenomena, such as birth, death, sex, etc. We struggle to understand them and to find our place in the cosmos and control over it. "These wishes call for the emergence of psychological force to come forth...to help human potentialities, psychological and otherwise, to crystalize and join the campaign to master these 'cosmic phenomena.' 'Spontaneity–which leads to creativity–is believed to be a psychological force which holds such requirements and serves such a purpose."

Teleological orientation: Kipper believes that spontaneity is a teleological concept which implies that "present activities are not the inevitable outcome of past determinants, but rather that the final course of events is their *purpose* (p 64.)"

Sui generis: Moreno never answered the question about the source of spontaneity. He wondered about the neurological system, or other constitutional endowment but this would still leave unanswered the question of spontaneity in the universe at large. One of his explanations is that spontaneity begets spontaneity, *sui generis*. (I will a suggest a source based on David Bohm's notion of the explicate order.)

Observable fact: Although Moreno wrote "But I postulated that spontaneity and creativity are observable facts and can be subjected to experiments, laboratory studies and systematic analysis. (1955, p 363,)" this claim is contrary to the idea that spontaneity is a form of energy. He also wrote "No one has ever seen spontaneity. Spontaneity is a hypothesis (p 372.)" No wonder people trying to grasp his conceptualization of spontaneity come away puzzled!

Trainable: Even before Moreno had worked out the Canon of Creativity, he was convinced that spontaneity could be increased through training. He writes "It was in the year 1923 that I set forth the dictum: 'spontaneity training is to be the main subject in the school of the future'(1946, p 130.)" this of course is the key to understanding and mastering creative process.

Adequacy and novelty: We are reminded of Moreno's favorite definition of spontaneity: "Spontaneity is the variable degree of adequate response to a situation of a variable degree of novelty. (1960, p 8.)" We also recognize that this conflicts with the idea that spontaneity is an energy. The response is the *carrier* of spontaneity in Kipper's terms. In any event, Moreno certainly associated novelty and adequacy with spontaneity.

Catalyzer of creativity: Again we quote Moreno: "Spontaneity and creativity are thus categories of a different order; creativity belongs to the categorizes of substance...spontaneity to the categories of catalyzer....(1953, p40.)

Kipper points out some of the contradictions between these characteristics of spontaneity and suggests that some of them actually better fit the concept of the warming up process, which is closely associated with spontaneity, defined as the operational expression of spontaneity. If we assign constitutional endowment, trainability, observable fact, and the qualitative criteria of novelty and adequacy to the warming up process, nothing contradictory is added to the latter and the warming up process may indeed be explicated.

Furthermore:

These suggestions leave the reader in the following position: "Spontaneity " remains in possession of its main characteristics, i.e., a non-conservable, non-accumulative energy," of a "cosmodynamic" origin, having a "teleological orientation" and serving as "catalyzer of creativity," (p. 73.)

The Spontaneity State

Moreno also described what he called the "spontaneity state." This is a condition in which the spontaneity of the individual is at a high level. It is the situation in which novel and adequate-to-the-situation actions are most likely to occur, and a state to which the creative artist aspires when he/she wishes to create. It is achieved through "an act of will" and a variety of techniques. It involves what we often call a state of inspiration, and it is often invoked through psychodramatic activity. The spontaneity state is probably one if what have come to be referred to as "altered states of consciousness," states which may be induced by hypnosis, meditation, or certain drugs.

The spontaneity state is equivalent to the natural "high" that Weil (1972) and others speak of, and is known by a number of names. Being "in the zone" is a term often used by athletics to indicate the high state of spontaneity that players strive to achieve, recognizing that the fullest expression of their talents and skills is most likely to occur when they are "in the zone." Being "on," "in the groove," and "in the flow" are other expressions commonly used to denote the experience of a high level of spontaneity. As a matter-of-fact, psychologist Csikszentmihalyi (1990) has published extensive investigations of the phenomenon he calls "flow" and describes as optimal experience.

Although spontaneity itself cannot be directly observed we often see its effects in other people's actions or experience a subjective sense of spontaneity. There are often strong subjective elements when one's own spontaneity is high. Because of this, Moreno was careful to point out that spontaneity is not itself an emotion. However, when one is spontaneous, the emotions that one is experiencing are experienced and expressed freely , and one's actions are congruent with the feeling.

There are a number of subjective indications when one's spontaneity is high. Being in a high state of spontaneity, one tends to experience a kind of harmony and integrity of emotion, thought, and coordination of action. One may experience a sense of being fully in control of oneself and capable of bringing all one's attention and energies to the situation at hand. A feeling of flexibility, a readiness to deal with anything that the situation may bring to bear, may also be experienced. Awareness of being in a high state of spontaneity is usually absent, coming only upon reflection at a later time.

A loss of the sense of time often goes along with the spontaneity state, a feature which also occurs frequently in so–called altered states of consciousness. As Kipper (1967) points out, there is an intimate connection between spontaneous behavior and the moment. An intense involvement in the moment dissolves the sense of past and future. One of the most common examples of spontaneous behavior is children's play. Most of us can recall the experience as a child of losing complete lack of time, and sometimes even of place, when we were deeply engaged in play with our friends, only being pulled back by the sound of mother's voice calling us to dinner. It is probably no wonder that Moreno got his first inklings of spontaneity when he was engaged with children in the Augarten of Vienna.

Spontaneous often seems to carry an implication of high energy and activity, even a lack of control. However, one reach a state of high spontaneity while being very quiet and physically inactive. Many people achieve the state through meditation or prayer. A quiet walk in a natural setting, in a forest or along a seashore may help others reach that state of internal harmony which signifies spontaneity. At the other extreme are athletes in an energetic game or match. Spontaneity can be a part of a full spectrum of activity.

While there is a whole range of emotions which can accompany high spontaneity, there is a category of emotions which are indicative of a lack of spontaneity. Moreno mentioned anxiety as the opposite of spontaneity and actually all of the negative emotions which are directed toward oneself, such as guilt and shame, are antithetical to the presence of spontaneity. When we are keenly aware of ourselves, of what we are doing, and especially of how we may look to others, we interfere with the functioning of our own spontaneity.

Summary

"I formulated this twin concept as the primary principles of existence in my earliest effort [*Das Testament des Vaters*] to comprehend the living universe in its entirety (1995, p 361.)" This statement bears repeating. It suggests, and I believe it is credible to assume that Moreno *intuited* what he eventually expressed as the Canon of Creativity. I think it is likely that it was through the mystical experience from which *Das Testament Des Vaters* emerged that he got his first inklings of how spontaneity and creativity were interrelated. This would make understandable the difficulties that Moreno had in giving verbal expression to a notion obtained in such a personal way. There is also the problem of presenting such a theory, one which essentially involves revealed knowledge, into a form that it can be widely shared by the scientific community.

Moreno described a procedure, spontaneity testing, by which he maintained that spontaneity could be measured empirically. Actually it would be more accurate to say that the spontaneity test measured spontaneous behavior from which spontaneity could be inferred. Despite his efforts there is little evidence of systematic research on spontaneity and it is likely that his suggested method had too many subjective factors to appeal to most researchers.

The concept of spontaneity is a radical one and Moreno knew it. He is proposing the existence of *a new form of energy which is unlike all known forms of energy*. It is unconservable energy. This is a startling proposition which flies in the face of a long-held and firmly entrenched physical proposition, the first law of thermodynamics, the law of conservation of energy. Moreno knew that it was an audacious suggestion in view of the fact that for years physicists have been searching for the unified theory which will integrate the known forms of physical energy. To propose that there is a form of energy which has not been here-to-fore discovered is an invitation to offhand dismissal by anyone who recognizes what Moreno is proposing.

However, Moreno was not unsophisticated and always optimistic that his theory of spontaneity–creativity was valid. He suggests that quantum research may eventually substantiate his ideas:

> Atomic nuclear research seems to confirm in principle, or at least does not contradict, the picture of the universe which the theory of spontaneity–creativity has envisaged. Its structure is not permanently set but when novel situations emerge, the responses to the surrounding field take the form of creative acts. As long as the universe was visualized as dominated by eternal, rigid laws, there was no place for "uniqueness" and for "explosive" changes and with it no place for creativity as the ultimate principle, at least not for the on-going, here-and-nowness of it. But a revolution has taken place on the highest level of conceptualization. We can say with greater certainty than ever that the supreme power ruling the world is Spontaneity–Creataivity. It has created a rational cosmos which coexists interdependently with man's perception of it but amenable to his intervention as long as he knows and abides by its rules (1955b, p 373.)

The Conserve and the Cultural Conserve

Moreno used the term *conserve* to signify the end product of a creative act. Anything that has been created, everything that can be named or identified is a conserve. The stars, the planets and other heavenly bodies are conserves, the results of the spontaneous–creative forces of the cosmos. Continents, oceans, and other geological features of this planet are also conserves. Every life form is a conserve, a biological conserve. Every individual member of every life form is also a conserve.

Moreno denoted the subset of conserves which are the result of human creativity as cultural conserves. Every language that exists, and every language that has ever existed, is a conserve, as is every word in each of them; so is every alphabet. All the tools, instruments, appliances and so forth that have ever been invented and fabricated, from the stone ax to the supercomputer, from the wheel to the space shuttle, are conserves, as are all the myths of all cultures and all of the religions with all their rituals, practices, hymns, sacred writings and the like. All the laws and social institutions; all of the artistic creations, stories and novels, drawings and paintings, musical compositions, dramas, poems, statues, etc.; all of the scientific theories and laws, and all of the technological inventions which are based on them; all of the great philosophical ideas, each and every one is a cultural conserve.

A singular characteristic of the cultural conserve is that once created, it can be reproduced any number of times without relying on the kind of spontaneity which is required for creating something original. Moreno often used the book as a prime example of the cultural conserve in this respect. Once the author has created the ideas and thoughts of which it is composed, and expressed them in words, the creative result can be replicated by the printing process any number of times. Once a computer has been invented, thousands upon thousands can be manufactured without having to recreate it. The same is true of all other conserves. Once created, many copies can be made of them through technological procedures (all technologies are cultural conserves, too, of course.)

Conserves can also be far less tangible, embodied by human beings. One of the more important conserves are the roles, complex organizations of behavior, in which we interact with each other. Other examples of this form of conserve are beliefs, myths, prejudices and attitudes.

Conserves give a distinct consistency and stability to the universe. In an ever-changing universe conserves provide a partial permanence which allows for a certain degree of predictability.

Without conserves the world as we know it is unimaginable. The universe would be a swirling mass of subatomic particles or energy without form or differentiation. Or perhaps it would resemble the end of the world that the law of entropy predicts, a uniform, undifferentiated, unmoving mass of atoms.

It was cultural conserves with which Moreno was most concerned and it is cultural conserves with which we need to be most interested. Cultural conserves make human culture and society possible. Moreno extols the positive side of conserves in these words:

> Conserves represent an enormous vitality and reality, not only in our human culture but in the universe at large. The universe is full of them. Every stone, plant and star, every animal organism is a conserve. Creativity is the universe itself, spontaneity is the key to its door and conserves are the furniture, the equipment which fills it. One cannot overemphasize the importance of the conserves which fill the storehouses of our culture....The intellectual conserves, as philosophical and scientific systems, are reprinted again and again, unchanged in every generation. Nothing new is added to them because their creators are dead, but they can still arouse other thinkers to new variations of these ideas or to new ideas. (1956, p 130.)

The conserve provides the basis for culture and society. The technological reproducibility means that the fruits of creative thinking can be shared with one's fellow men. Only one person needs to invent the wheel for all to make use of it. Some conserves are the result of group creativity as language must have been. In current times most significant advances in science and technology are the products of teams. Watson and Crick and their discovery of the double helix structure of the gene come to mind, or Bohr, Heisenberg, and Schrödenger and quantum theory are examples. That which one generation creates, whether language, mythology, scientific fact, or tools and instruments, can be passed down to their children. Whether the brainstorm of a single individual or the collaborative genius of a group, the works of the creative intelligence of one era are available to members of future generations. The cultural conserve gives humankind and human society its distinctive characteristics.

The conserve is of major importance in the creative cycle. Conserves inevitably figure into new creative acts. A great, new novel is expressed in words which are of course conserves. Words, in turn, are composed of the letters of the alphabet, another conserve. The ideas may be new and fresh and never put together in quite the same way, but the language is indispensable for the ideas to be communicated. New theories are expressed in the conserve of language, too while musical composition makes use of already invented musical notation. An artist creates a master painting making use of previously developed brushes, pigments and other materials. Scientists invent tools and then use them to test their theories. It is impossible to imagine a creative act which does not depend upon cultural conserves.

There is a complex relationship between cultural conserves and spontaneity. For a musician to perform a piece of music, one which has been performed a number of times, or for actors to perform a play after the first time, requires that they infuse the performance with spontaneity or run the risk that the performance will seem dull and not alive. The same is true for a familiar prayer. Each recitation requires that the individual invest it with spontaneity or it simply becomes words with little meaning to the individual. Thus spontaneity gives a freshness to conserves without which the conserve becomes mechanical and lacking in liveliness.

On the other side, conserves may serve to trigger spontaneity. A stirring speech, such as Franklin Roosevelt's address to congress on December 8, 1941, may serve to motivate and activate a whole nation to the unwelcome task of going to war on another continent. The crisis papers of Thomas Paine are given credit for playing a part in stimulating the American revolution. A piece of music, a painting, a novel, an essay, an instrument, a theory–any conserve may serve to inspire others to creative acts.

Paradoxically, the cultural conserve not only is the vehicle for culture and society, but is the source of much of the dissension and violence in the world, and, Moreno feared, could contain the germ of potential destruction of the human race. Beyond the obvious fact that science and technology have deciphered secrets of atomic structure, giving us a potential means for destroying most if not all mankind and simultaneously making the planet an impossible habitat, are intellectual conserves which can be destructive in more subtle ways. Such conserves include religious convictions and beliefs, political philosophy, nationalism and racial prejudice, and scientific theories, among others. Awareness of religious wars and the inquisition is too widely known to belabor here. It is also pretty obvious that the other great cause for war has been due to differences in political philosophy.

It must also be obvious how racial and ethnic prejudice, as well as nationalism are conserves, deeply entrenched notions and patterns of thinking that serve destructive ends. The events in the last decade of the 20th century in Yugoslavia, a federation of six republics and multiple nationalities which had been cobbled together at the end of World War I, offer an example. Under the totalitarian Communist regime the various ethnic groups were able to function in relative cooperation and intergroup dissensions remained repressed. Economic problems combined with the loss of the leadership of Marshal Tito in 1980 resulted in the re-emergence of national and ethnic discords. This conflicted with the attempts of Serbia to maintain and dominate the federation, resulting in war, first in Slovinia, then Croatia, and finally Bosnia.

.   The fact that the hostilities between those national and ethnic groups can be traced back at least a thousand years testifies that the underlying intellectual conserves, convictions of the members of one group that those of another are unworthy, untrustworthy, dangerous, or in the case where a race is enslaved, not fully human, are incredibly tenacious. Moreno was convinced that sociometric investigation would enlighten us about how these conserves are maintained as well as bring about de-conservation or dissolving them.

How scientific theories can function as a negative force as well as a positive one is worthy of further discussion. Thomas Kuhn (1962) has disabused us of the notion that science is a slow and steady process of accumulation of knowledge in and about a particular field of inquiry. He points out that scientists work within a certain framework of deeply imbedded beliefs, a paradigm, shared by the scientific community. A paradigm is maintained until it is no longer consistent with the results of scientific investigation. At that point, there is a revolution and features of the existing paradigm are discarded and a new one within which scientific investigation continues, takes its place. This is not a smooth and easy process, however, and paradigms are not easily overthrown. Scientists tend to hold tenaciously to their beliefs, inculcated during their education in their particular field of study. A major example of paradigmatic revolution occurred with Einstein's theory of relativity which challenged the notion of an absolute space and time, concepts which had been uncontested since the time of Isaac Newton's great works.

These beliefs, which really are theories, are conserves. They are conserves which are so highly valued that they often are regarded as inviolable truths, and are generally safe from inspection and reflection. Therefore, a scientist doesn't often examine or even discover these beliefs. One such belief is that scientific work is objective. A common belief is that the scientist asks the question, designs the research which might answer it and accepts without subjective involvement the result that is obtained. Actually the scientist is always subjectively involved, beginning with the choice of the problems he/she chooses to investigate to the interpretation of results. Kuhn believes that these choices and interpretations are severely constrained by the prevailing paradigm.

In short, those conserves which Kuhn labels paradigms result in a narrowing and rigidity of thought on the part of scientists. New ideas, even unanticipated results are often ignored and discarded. Theories which diverge too far from paradigmatic beliefs are considered as not worth considering. It is quite possible that among the thoughts, ideas and theories that have been overlooked and ignored because they lie outside our paradigms are some which could alleviate much of the collective misery which humankind faces today.

Moreno (1953) anticipated some of Kuhn's ideas in a discussion of The Emergence of the Scientific Method (p 23.) The real advances in science, he wrote, come from the insights of a few geniuses. The job of the scientist is to "steal" the ideas of the genius and make them understandable to all the non-geniuses in the world. He uses the myth of Prometheus, the demi-god, who stole fire from Zeus and brought it to mankind.

I will argue at a later time that Moreno's work was too much at variance with the prevailing paradigms of the social sciences of his time, and that this is a major reason for lack of acceptance of his philosophy and theories, even while his methods were widely adopted.

The problems which conserves get us into are not within the conserves themselves. The problem is that most of us prefer to live a conserved life rather than a spontaneous one. We are all dominated by conserves, Moreno says. The more we are dominated by conserves, the greater is the probability that our behavior is consistent and predictable. If It weren't for the spontaneity which re-energizes them, renews them, much of the world would be deadened. Society would function like a machine and we would function like robots.

Spontaneity is the principle of "unconservability" and "unpredictability". Conserve is the principle of conservability", "constancy" and "predictability". Creativity can remain unfrozen and linked to spontaneity, or it can freeze and be linked to conserves. (p 132.) Perhaps it is a genetic or constitutional predilection, but it seems as if most people are looking for answers, the right way to do things, the right way to behave, how to be. The struggle is between living spontaneously versus living according to conserves. There is no question that today's society encourages us to live comport ourselves by fairly highly prescribed rules, although many who live spontaneously engender approval and envy from fellow men. Spontaneity is an appreciated value and those who are spontaneous are often perceived as more real, more trustworthy than those whose actions are calculated and planned.

Lewis Yablonsky, sociologist, close friend, student and colleague of J. L. Moreno, wrote about the dangers of the over-conserved life style, coining the term *robopathology*, (1972.) Yablonsky's worry is not so much focused on the rebellion of the robots as it is about the dehumanizing effects of a technological society:

> The problem of the physical machine takeover of the destiny of people is obviously a phenomenon of enormous proportion. An even greater problem, one that is more subtle and insidious, exists. This involves the growing dehumanization of people to the point where they have become the walking dead. This dehumanized level of existence places people in roles where they are actors mouthing irrelevant platitudes, experiencing programmed emotions with little or no compassion or sympathy for other people (p. 6).

Yablonsky calls this condition robopathology. It refers to people who function more like robots than like compassionate human beings. Robopathology, he suggests, is the dominant socio-psychological pathology of our era and is a product of technocracy, which imposes excessive planning and encourages conformity to social values and standards. Technocracy represents another way in which we are endangered by the conserves of our own creation.

In my opinion, the notion of the conserve is one of Moreno's most important contribution to science. It is a single concept, a category which includes both tangible artefacts and intangible entities like hypotheses, theories, and inspirations. Conserve can be thought of as cutting the Gordian knot of Cartesian mind-body dualism. The tangible body and the intangible thought are actually on a continuum. A product of thought, a scientific theory for example, or an idea for a new mousetrap, an invention, is at first a subjective event in the mind of its author. It is a conserve because it can be replicated in a number of ways. The theory can be articulated by lecture, or by an article in a scientific journal. Plans for the mousetrap can be drawn or an actual version of it produced. The published theory is read by others in whom it stimulates thoughts both for and against it. Hypotheses, more conserves, are generated to test its validity.

The Warming Up Process

The warming up process has already been defined as the operational manifestation of spontaneity. The warming up process is of special importance because it is the aspect of the creative process over which we have some control. By learning about one's own process of warming up, one can increase the possibilities of responding to situations spontaneously. Through psychodrama, protagonists can not only free themselves from conserves but can also explore their process of warming up to problematic emotions, relationships, conserves or situations which interfere with creative living. Identifying and altering those factors which tend to block the warming up process promotes greater creativity. In short, one can actually learn to live one's life more spontaneously, more creatively.

Human behavior is process. There is a beginning, a continuing, and a termination of every act and every event. Any complex act, especially an act of creating, begins with preparation, preliminary actions that serve as preparation. This is, in Moreno's terms, the warming up process:

> The warming up process, the operational manifestation of spontaneity, is a general condition existing before and in the course of any creative act--before and during an act of sleeping, eating, sexual intercourse, walking, artistic creation or any act of self-realization. Spontaneity is generated in action whenever an organism is found in the process of warming-up. (1951. p111.)

Those who engage in creative activities, such as writers, artists, performers, and so forth are usually very sensitive to their warming up processes. It is not unusual that they evolve preparatory rituals as a way of preparing themselves to engage in the creative activity. Just as athletes "warm" themselves up physically and mentally before engaging in an athletic event, creative artists "warm up" to the task in a purposeful, systematic way, knowing that success may depend upon how well they accomplish this preliminary step. Moreno studied the warming up processes of the players of the Stegreiftheater. Here is how he described it:

> The Impromptu agent, poet, actor, musician, painter finds his point of departure not outside, but within himself, in the spontaneity "state." This is not something permanent, not set and rigid as written words or melodies are, but fluent, rhythmic fluency, rising and falling, growing and fading like living acts and still different from life. It is the state of production, the essential principle of all creative experience. It is not given like words or colors. It is not conserved, or registered. The impromptu artist must warm up, he must make it climbing up

the hill. Once he runs up the road to the "state," it develops in full power. (1944, p 44.)

Warming up may involve a myriad of circumstances. For example, the warm up of a writer to a masterpiece may well involve all the activities through which he/she developed writing skills, as well as all the works that were created before the masterpiece is begun. The warming up process may also involve all the learning and reading about subject matter included in the novel. It may involve life experiences which were rich with meaning, and which the writer felt strongly about reporting in a fictional form. It may involve inspiration from any number of sources such as poems, other novels, a piece of music heard, a statue or painting observed, a relationship, and so forth. It may involve developing and practicing skills need for the creative work. Any experience may be a part of the warming up process for a subsequent act–and in some sense is.

Characteristic of warming up is a state of tension, the result of a demand either self-imposed or from a situation in which an one finds oneself. Life, Moreno was fond of pointing out, is full of surprises. It is through our spontaneity that we deal with the unexpected, that we "rise to the occasion." Moreno devised a "spontaneity test" which consisted of putting the protagonist into a situation which demanded action. For example the protagonist might be asked to consider himself at home when he smells smoke and realizes that the house may be on fire. As he responds to the situation, additional conditions are assigned. Perhaps first the protagonist rushes to the phone to call the fire department. As he does so, the director informs him: "The baby is upstairs in the nursery and your mother-in-law is in the back bedroom. As the protagonist responds to these suggestions, new ones are offered, i.e., "Your wife's valuable jewelry is in your bedroom," and so forth.

Although Moreno called it a test, the spontaneity test might better be considered spontaneity training. There is actually no objective measure for the test. By and large, the degree of spontaneity is assessed by the consensus of an audience who discuss both the novelty and appropriateness of the protagonist's performance. Sometimes in a training

132

group, a director and an auxiliary devise a "test" situation and group members, one at a time, enter and respond to it. An example of such a situation might be that the trainee is asked the name of a best friend. The auxiliary is introduced as an acquaintance whom the protagonist runs into at a party. After a moment of chit-chat, the auxiliary says, "Did you hear that [name of best friend] was just arrested for drug dealing?"

The higher levels of creativity such as composing a great symphony, writing a successful novel, making a great painting or sculpture, or conducting meaningful scientific explorations is seldom the result of inspiration (or spontaneity) alone. They usually involve hard work which both precedes and follows the inspiration. Moreno reminds of us of Thomas Edison's statement that "Genius is one percent inspiration and 99 percent perspiration" and says that perspiration is the folkword for "warming up." He notes that warming up does not necessarily achieve the desired end of the spontaneity, the creation of a great work of art or a magnificent new scientific theory.

We have previously mentioned the debate between Pitirim Sorokin and Moreno, published in Sociometry, Vol. XVIII. First comes Moreno's article on spontaneity–creativity, followed by Sorokins critique. It is to Moreno's reply that we now turn. Moreno not only clarified further his ideas about spontaneity–creativity, but he described in detail the warming up process which led to his decision to write the reply. It illustrates the warming up process better than any of his other writings. The article begins:

> Creativity is *the* problem of the universe; it is, therefore, *the* problem of all existence, *the* problem of every religion, science, *the* problem of psychology, sociometry and human relations. But creativity is not a "separate" mystic, aristocratic, aesthetic or theological category; if it is on top, it is also on the bottom; it is everywhere; if it is in the macrocosmos, it is also in the microcosmos; if it is in the largest, it is also in the smallest; in the physical and social atom, it permeates the common and the trivial; it is in the eternal and the most transitory forms of existence; it operates in the here-and-now, in this pencil and in this paper, as I am writing these words to the reader.

Then he asks: "How did the writing of these words come about?" and proceeds to give an account of his warm up to the writing of this paper:

It came about because I have before me the comments of Sorokin to my "system of spontaneity–creativity." What I have before me, however, is not Sorokin himself talking. I was not present when he wrote his article in *statu nascendi*. What I have before me is already a finished product, a conserve. Here it is on my desk; it was on Sorokin's desk two days ago but how it is here. If it would have been multiplied in thousands of copies, it would remain the same–unchanged–a conserve. Multiplications do not make a conserve less or more of a conserve; it is in cold storage, dormant energy. I could put it aside and file it, judging that it is unworthy of publication or that it should be published at another time; but as I read it, I got warmed up and was inspired to answer him. What happened–I am trying to give a "phenomenological" description of the process–is the following: as I read it, two things happen. One is the reproduction of Sorokin's article in my mind. I lift it frm the paper upon which it is typewritten, so to speak, in order to duplicate it within my own thinking. At this point Sorokin cannot "add" anything to it, however many times I read it, although it may become "clearer" to me what is expressed in the conserve. But the repetition of reading does not extend the content–and here a second phenomenon enters into the picture–as I begin to react to it. I get, as we say, "warmed up"; all kind of ideas come to me: "we agree on this point–creativity"; "Oh, this is not what I mean by spontaneity"; "Most people who have read my many books and articles on creativity and spontaneity will immediately realize the that the difference is not serous enough, it is a matter of semantics"; "Here is a brilliant suggestion about the study of non-spontaneous factors". But suddenly I read a sentence about creative "conserves" which alarms me. Conserves by themselves without an "intervening"factor do not become creative. We build up the conserves and try to make them look like idols. I see the ghosts of Plato and Aristotle coming back. It makes creativity look stale and makes a puppet out of God. No! No! Aristotle and Plato have idolized the conserve like hundreds of other philosophers of Western civilization, including Spinoza from whom God became a huge existential conserve, similar to St. John's logo for whom "The Word" implies an eternal, unchangeable conserve; here he was betraying Jesus who was intuitively set against it. More than the creators, we admire their conserves until they become sacred, like the "Bible", Beethoven's music or the "Pyramids."

At this point, I come to *the decision that I must answer it*; and herewith I answer Sorokin's reply to my theory of creativity. It is a new phase in my warming up process which takes place. Up to this pint, I am warmed up to Sorokin's comments, just reacting to them without any further plans in mind;

but now that I have decided to answer the comments, my warming up is directed towards a goal of my own–the production of my (this) article. A number of extraneous hindrances enter the situation: my secretary has a cold and may not be able to work for a few days; there is a printer's deadline for the volume; I am scheduled for a lecture tour to start a week hence; the consequence is that I have to rush or cancel my lecture tour, get a substitute for my secretary or try to delay the deadline. Then there are a number of internal hindrances in my production: if I answer Sorokin, I should give him a chance for another rebuttal;–a certain inertia sets in–I feel tired, I have been overworked and should take a rest;–my wife steps in and says that we should go to Florida for two weeks.–but these various extraneous and internal autobiographic factors, including the coercive ones, are *part* of my warming up process which is the operational manifestation of spontaneity. The sum of it is that I am gradually getting "ready" to write. I am only uncertain as to *what* I should say beyond what I have said so many times, I am at a loss as to what *form* it should take in order to put my ideas across better. In this moment of indecision, I have an inspiration: *to start my response to Sorokin's comments in the manner in which I am doing it "now"* by giving an actual account of what I am going through, convinced that by such an existential mode of presentation, I can make more plausible to Sorokin and to the readers my concept of the relationships between creativity, spontaneity and the conserves. This inspiration sets me off to create. It is the intervening factor, the climax of my spontaneity in this particular situation. (1955c, pp 382-384)

Here we have the written equivalent of a psychodramatic soliloquy with Moreno documenting his warm up to the writing of a paper as he wrote it.

Concept of the Moment

Moreno continuously emphasized the here-and-now, the moment. Spontaneity exists only in the moment. But what are we to make of the moment? Obviously it has to do with time. Is it nothing but a small slice of time, a "fleeting transition between past and future, without real substance; that it is intangible and unstable and therefore an unsatisfactory basis for a system of theoretical and practical philosophy" (1946, p. 104)? And if so, how small a slice, a second, a minute, an hour, a day, a nanosecond?

Moreno considered that he had made a significant philosophical contribution in connecting the concept of the conserve to that of the

135

moment. The moment involves a subjective experience, an experience in which a change in the situation takes place. To be experienced as a moment, a change that can be perceived and responded to by the subject must occur that serves to emphasize an event as "separated from past and future moments as one particular moment." This change is signified by the emergence of a conserve.

Bergson, whom Moreno identifies as a significant influence upon his own thinking, came close, Moreno says, to the overall problem of creativity. Bergson considered time to be constant and ceaseless creative change and believed that it could be directly experienced as such. His concepts of *elan vital* and *durée* carried that implication. However such a world would be without conserves and would be experienced as a kaleidoscope, always moving and changing without any consistency.

Everything exists in the moment, in the here-and-now, including the past and the future. The past is our here-and-now perception of previously experienced moments; the future is our anticipation in the here-and-now of upcoming events. This principle holds in the psychodrama. A protagonist may be re-enacting a past event but the psychodramatization of that event is occurring in the here-and-now, the moment of the psychodrama, and the protagonist is expected to respond accordingly.

In terms of human-made time, the moment is equal to the lifetime of the conserve which defines it. In that sense, my life is a moment. It also subsumes innumerable other moments which represent the meaningful events in my life. David Bohm (Bohm & Peat, 2000) has a similar concept. The implicate order, in his system, is timeless. Time is a part of the explicate order and is organized by society.

Summary

Creative change seems to be a benchmark of our universe, and humankind has shown a considerable precociousness toward certain aspects of creating, primarily in the domain of dealing with our physical environment. Moreno's concern is that humankind has been prodigiously creative in the realm of science and technology but has had a tendency to overvalue the *products* of that creativity. In our admiration of what we have created, we have neglected to master the *process* of creating. We have thereby put all of humankind in jeopardy. We have created the means by which we can destroy much if not all of humankind, possibly life itself as we know it, but at the same time we seem unable to create a society in which we can peacefully and in collaboration with our fellow human beings.

While a solution to these problem of living was his goal, Moreno realized that no one person, no two people, no small group of people could bring it about. This was a task so immense that it required the cooperation of everybody, every living person. And *that* is the meaning of his oft-quoted and, he sometimes said, little understood opening statement in *Who Shall Survive?*: A truly therapeutic procedure cannot have less an objective than the whole of mankind.

References

Aculino, J. (1954) Critique of moreno's spontaneity theory. *Group Psychotherapy, VII.* 148-158.

Csikszentmihalyi, M. (1990) *Flow the psychology of optimal experience.* New York: Harper & Rowe.

Kipper, D. (1967) Spontaneity and the warming-up process in a new light. *Group Psychotherapy,* XX, p.62-73.

Moreno, J. L. (1923). *Das Stegreiftheater. (The theater of spontaneity)*, Berlin/Potsdam: Kiepenheuer.

Moreno, J. L. (1934). *Who shall survive?: A new approach to the problem of human interrelations.* Washington, DC: Nervous & Mental Disease Publishing.

Moreno, J. L. (1946). *Psychodrama, vol.1.* Beacon, NY: Beacon House.

Moreno J. (1947) *The Theater of Spontaneity.* New York: Beacon House.

Moreno, J. (1951) *Sociometry, experimental method and the science of society.* Beacon, N.Y.: Beacon House.

Moreno, Jacob L. (1953). *Who shall survive? Foundations of sociometry, group psychotherapy and Psychodrama*, Beacon, NY: Beacon House.

Moreno, J. L. (1955a) Canon of creativity. *Sociometry,* XVIII, 359-360

Moreno, J. L. (1955b) Theory of spontaneity and creativity. *Sociometry*, XVIII, 361-374

Moreno, J. L. (1955c) System of spontaneity–creativity–conserve, a reply to P. Sorokin. *Sociometry*, XVIII, 382-392.

Moreno, J. L. (1960) Spontaneity–Creativity–Cultural Conserves (pp. 8-14), in Moreno et al, (eds.) *The Sociometry Reader*. Glencoe, Ill.: The Free Press.

Moreno, J. L. (1971) *The words of the father*. Beacon, N.Y.: Beacon House.

Moreno, J. and Moreno, F. (1944) Spontaneity theory of child development. *Sociometry*, VII, pp. 89-128.

Sorokin, P. (1949) Concept, tests, and energy of spontaneity–creativity. *Sociometry*, XII, 215-224

Sorokin, P. (1955) Remarks on J. L. Moreno's "theory of spontaneity–creativity". *sociometry*, XVIII, 374-382

Weil, A. (1972) *The natural mind, a new way of looking at drugs and the higher consciousness*. Boston: Houghton Mifflin.

Yablonsky, L. (1972) *Robopaths*. Indianapolis: Bobbs-Merrill Company, Inc.

STRATEGIES OF DIRECTING

"A plan, method, or series of stratagems for obtaining a specific goal or result." Def 4 for "strategy", Random House Dictionary of the English Language, 2nd Ed., Unabridged.

Most experienced psychodrama directors have developed strategies, "plans of action," which they use, either consciously or without the awareness that they are using them. It would seem preferable to be aware of one's strategy. A strategy is a conserve, of course, and as such may be utilized in the service of creativity—or as a substitute for spontaneity. In keeping with spontaneity-creativity theory, we should attempt to cultivate the former and to avoid the latter. Beginning psychodrama directors generally have their hands full tending to the protagonist's warming up process and perhaps their own, deciding which techniques can best be utilized in the protagonist's behalf, being aware of the rest of the group, and all the other things to which a beginning director must pay attention. Thinking in terms of strategies of directing should probably not be attempted until after one has mastered the rudiments of psychodrama and has become rather familiar with the method, although it cannot hurt to know that there is such a thing as a strategy and to have an idea of what one is.

The concept of strategies of directing developed from my observations of experienced directors during the years when I was first learning the method. Watching the stars of the psychodramatic collective of those days, directors including Zerka Moreno, Jim Enneis, Martin Haskell, Hannah Weiner and others, I was puzzled by the big differences in the manner in which each went about directing a psychodrama. Yet each drama was remarkably effective. I was frustrated because I was trying to learn the correct way to direct. Although J. L. Moreno emphasized that directors of psychodrama develop their individual styles, I thought there was something more involved.

When I began to train others, I expected a trainee to direct the drama exactly as I would have, had I been in the director's role. I soon learned that my way wasn't the only way, that there seldom is "*the* right way" but that there may be a number of right ways, some better than others, perhaps, but successful none the less. There are of course some wrong ways.

As I puzzled on these matters, it gradually occurred to me that what I was attending to were some significantly different ways in which directors formulated and approached the problems presented by their protagonists. Here is where the life experiences of the director interact with the life experiences of the protagonist.

Although the notion of a strategy may seem to be inconsistent in the context of spontaneity methods, this is not really so. Though neither the protagonist nor the director know at the beginning of drama just what will emerge or how the drama will turn out, there are elements of planning and organization which the director utilizes in the task of co-producing the psychodrama. These include some of the things I am calling strategies of directing.

A strategy will be influenced by the director's understanding and perception of psychodrama as a method, and how one perceives its use as a therapeutic instrument. Also involved is the director's philosophy and understanding of human nature, as well as one's understanding or theories of psychopathology and emotional disequilibrium. Most directors seem to be able to utilize several strategies in their work, moving from one to another in response to the protagonist and the problems which the protagonist presents. I have identified several strategies which appear to be fairly common within the psychodrama collective. I consider these to be major strategies with an infinite number of variations. I suspect that there are other major strategies and probably a lot of variations on each one. A strategy is not identical to personal style, but the director's style is certainly influenced by the particular strategies that the director tends to use frequently.

Four strategies of directing are described in the remainder of this paper Each of these has been observed in the work of certified psychodrama trainers and nationally recognized leaders in the fields. Many of the observations from which this concept is derived were made at major psychodrama conferences. I am calling them 1) the psychodynamic strategy; 2) the sociometric strategy; 3) the psychodrama tic strategy; and 4) the sociodramatic strategy.

The Psychodynamic Strategy

This is an approach which many of us who were originally trained in traditional forms of psychotherapy naturally gravitate toward and which we probably utilize initially until we get familiar with all the possibilities that psychodrama offers. This strategy tends to emphasize the therapist role of the director and the patient role of the protagonist in a more traditional manner. The protagonist is perceived as having a problem, often perceived as an internal conflict, and the director is perceived as an expert who will help him identify and resolve it. Further, the problem will likely be construed in terms of the protagonist's internal dynamics and conflicts. The director-therapist's job is to help the protagonist identify and resolve these internal conflicts which the protagonist may be projecting upon others or his environment.)

The director who relies heavily upon this strategy will often direct dramas in which there is a lot of intrapersonal psychodrama. That is, he will have the protagonist confronting or encountering parts of self, his anger, his fear, his sadness, etc. He may frequently use the double ego technique and multiples doubles in which an auxiliary is assigned to each of several internal voices which are giving the protagonist contradictory advice in a dramatized situation.

Of course there are many other systems of internal dynamics besides the psychoanalytic one which is usually what the term psychodynamic refers to. However, if the director using this strategy has been trained in transactional analysis, he will produce the parent, adult, and child ego states and examine the problem from this perspective. If an encounter is involved, he may produce the ego states of both parties. The more sophisticated psychodramatist-sociometrician will cast these issues as role conflicts, and will deal with role-rigidity or over-conserved roles, and may end up in a role training situation.

The main characteristic of this strategy is its focus on fixing something internal, something within the protagonist which is interfering in his living his life easily.

The Sociometric Strategy

This is a social systems approach, and begins to be incorporated by many new directors fairly early in their directing, as they become acquainted with the sociometric aspects of Morenean theory. The director whose style incorporates this strategy and emphasizes this approach will attend a great deal to the social environment of the protagonist, the protagonist's family, either current or original. If, as frequently happens, the protagonist produces a scene in his home, current or past, this director will find out exactly who else is living here, assign these roles to auxiliaries, and get some degree of the family dynamics into the action. If the scene is at the office, we will learn something of the protagonist's work social atom.

This director also tends to put as more emphasis upon inter-personal conflict then upon intra-personal conflict. Social investigation is a frequent activity of this director, carried out through role-reversal, use of auxiliaries, and the directors directions. He often gets the protagonist to define his goal as resolving inter-personal conflict through a greater understanding of 1) the other person; and 2) the protagonist's sensitivities from earlier

142

experiences. The main characteristic of this strategy is producing sociometric data in the drama. This director loves dinner scenes!

The Psychodramatic Strategy

The director who employs this strategy perceives his job as helping protagonists tell their stories, the stories of their lives. These directors tend to minimize the therapist role of the director and maximize the role of dramaturge. Rather than focusing on either the internal or social dynamics of the protagonist, this strategy calls for assisting the protagonist to express himself in dramatic fashion. The psychodrama should make apparent to all who take part in it that the protagonist is showing the world that "This is what it is like to be me, to live my life."

These directors will want to produce scenes which tell the protagonist's story. They will be looking initially for a beginning of the story. Although some stories can be told in a single scene, many times these directors will produce several. They will also be encouraging protagonists to produce not only what may have happened before, but to expand upon the original, to express the feelings which were not or could not be expressed in the real-life experience. They will encourage expansion from the roles of the significant others in the scene as well, thus engaging social investigation as does the sociometric director. When internal conflicts are a part of the story, and they often are, these directors will produce the intrapersonal dynamics. Often this stratagem will take the drama to earlier stories, experiences in which one or another of the conflictual factors was established. However the psycho dramatic director will always anchor intrapersonal drama in an an interpersonal scene.

Once the protagonist has told the story as perceived by the protagonist, this director may look for injustices and incompletions. These are experienced by both the protagonist and the group as tensions. Resolution is found through the use of surplus reality to complete incompletions, and to correct injustice. Often it can be put into a formula of: "This is what happened" followed by "This is what should have happened."

The Sociodramatic Strategy

This approach is marked by the director's interest in the collective aspects of the roles that emerge in the drama. Thus, if the protagonist wants to deal with the relationship between himself and his wife, the director will want to produce dramatically information about the collective roles of wives and husbands in the protagonist's culture or subculture. He tends to give auxiliaries a lot of freedom in the way that they create their roles, encouraging them to incorporate their understanding of the role into their portrayal. At times, the protagonist may appear either confused or resigned by all the activity that is going on around him, fueled by the spontaneity of his auxiliaries. Something akin to the following dialogue often seems to take place in a drama directed through this strategy:

Protagonist: (to wife) I just have had it with you. I don't like to come home after a hard day's work and find that you've got plans to go out in the evening on your own. I don't know where you are going and I don't know who you are hanging out with, and I just won't have it!

Auxiliary Wife: Well, you won't ever take me out! I don't know why I have to sit around at home with you when you are so tired you don't want to do anything but watch T.V. and go to sleep! You don't own me! I can go anywhere I want anytime I want!

Protagonist: (to Director) That's not right. She never says that. She is always apologetic and sorry, and—

Director: You don't know everything about your wife! Maybe she doesn't say that but maybe that is what she is thinking! Go ahead and respond to what she just said.

The director with a psychodramatic strategy would be more apt to call for a role reversal and ask the protagonist to demonstrate what he wife would say and do.

Like sociodramas, these psychodramas sometime stir up more energy and questions then they resolve. On occasion, auxiliaries get overheated and essentially take the dramatic focus away from the protagonist. The participants may come away from the session quite stirred up by the experience and possibly with some harsh feelings toward the director.

Comments about the Uses of These Strategies

Each of the approaches outlined above is a legitimate way to go and can lead to an effective psychodrama. Each is especially effective under certain conditions. The first approach, the one I have called psychodynamic, seems to me to be too limited to use by itself except with protagonists who are obsessively focused on their thinking. It does not make good use of the fullness and richness of experience of which psychodrama is capable. On the other hand, as an adjunct to other strategies, it can be extremely helpful and should certainly not be avoided. A good indication for its application is when the protagonist is resistive in the psychodramatic sense. That is, the protagonist finds it difficult or impossible to carry out the director's directions. However, the beginning director (and the advanced one, too, for that matter) will never go wrong by making sure that intrapersonal scenes or productions are anchored at both ends in interpersonal situations.

A caveat: intrapersonal psychodrama may be less easy for the group members, especially those who don't get to be auxiliaries, to warm up to and get deeply involved in.

The director who utilizes the sociometric strategy will very likely help the protagonist generate much richer information than one who sticks to intrapersonal things. Probably all problems and conflicts have both an intrapersonal aspect and an interpersonal aspect. They also have sociodramatic as well as psychodramatic elements. The protagonists problems and issues can be explored and worked on from either or all sides. It is generally easier for us to look at the part we play in interpersonal difficulties after we have had a chance to blame the other fellow for it, and to give expression to the anger, hurt, or disgust we may feel. At this point, most of us are more ready to do the role-reversal and look at the situation from the other's point of view—which also means to look at ourselves from a different perspective. Although either intrapersonal or interpersonal encounters can lead the protagonist back to the earlier experiences from which factors may be intruding into his present day business, it seems that the interpersonal approach may be more facilitative of this maneuver.

The sociometric strategy has the advantage of sensitizing both the protagonist and the group to the extent to which we are responding to the pushes and pulls of the social systems of which we are a part. This awareness can be utilized to free up certain behaviors making it possible for us to act in more comfortable ways and to maximize our influence upon our social systems.

There seem to be a number of advantages of the dramatic or story telling approach. First of all, it never excludes excursions into intrapersonal or sociometric realms while providing a natural setting or framework within which to examine these conflicts. Secondly, it is a very natural way to develop a drama, and is consistent with the goals of a humanistically oriented psychotherapeutic approach. I quote Sheldon Kopp:

> Each man's identity is an emergent of the myths, ritual, and corporate legends of his culture, compounded with the epic of his own personal history. In either case, it is the compelling power of the storytelling that distinguishes men from

146

beasts. The paradoxical interstice of vulnerability, which makes a man most human rests on his knowing who he is right now, because he can remember who he has been, and because he knows who he hopes to become. All this comes from of the wonder of hs being able to tell his tale, (page 14.)

And Kopp quotes Eli Wiesel: "God made man because he loves stories."

I believe that when we use this telling-the-story principle in our directing, we accomplish several things. One is that we all have a need to tell our stories related to the existential need to know that our life is worthwhile, that it has made a difference that we have lived. We need to let the world know how we have experienced it. In the telling of our story, we may also free ourselves of some parts of the past which are burdensome to us. This is a point which Kopp emphasizes. And finally, but far from least important, I think that a psychodrama which tells the protagonist's story communicates more deeply to the members of the group, both those who take auxiliary roles and those who simply witness, representing world opinion, as Moreno once described the audience. That is to say, the protagonist's story communicates to deeper levels of us, stirs us nearer the central parts of the self, than does psychodynamic or sociometric material. The latter speaks more loudly to the intellectual parts of us, the curious, problem-solving aspects of us.

Telling-the-story also fits with the fact that psychodrama is above all an art form. Psychodrama literally makes the protagonist the hero/heroine of his/her own life. It reveals both the uniqueness of the protagonist and his/her commonness with others, and in sharing, the other members of the group have a chance to experience their respective uniquenesses and likenesses. It is the approach which stands to make the best use of infinitely rich potentials of psychodrama.

147

REFERENCES

Kopp, S. 1972 *If You Meet Buddah on the Road, Kill Him!* Science and Behavior Books, Inc.:Palo Alto.

PSYCHODRAMATIC PRODUCTION OF DREAMS:
"THE END OF THE ROAD"

J. L. Moreno to Sigmund Freud (1912): "You analyze their dreams. I try to give them the courage to dream again."—*Psychodrama, Vol. I,* page 6.

Introduction

From prehistoric times, man has apparently been fascinated by his own dreams. Originally this preoccupation was focused on attempts to predict future events from dreams, or, in the earlier days of the "period of enlightenment," with the debunking of such notions. Sigmund Freud, declaring the dream the "royal road to the unconscious" shifted the emphasis from the future to the past and sought understanding of his patients' deeper, darker impulses through analysis of their dreams. To the psychodramatist, dreaming is a creative process and the dream can be thought of as representing the early stirrings of creative impulses. As such the dream may both reflect past experiences as well as point to future directions in the dreamer's life.

Dreams that eventually become psychodramatized tend to possess one or more of three characteristics. They are nightmares, they are repetitive, or they seem important but puzzling to the dreamer. The goal of psychodrama dream production is to "extend the dream beyond the end which nature has set for the sleeper or at least the end which he remembers" (Moreno, J. L., 1951). In doing this, the protagonist may be assisted 1) in confronting the fears underlying the nightmare; 2) in working his way through resistances to finishing the processes underlying the repetitive dream, or; 3) in exploring the meaning of the perplexing dream.

A Dream Psychodrama

The dream presented for this discussion illustrates the classical procedure for psychod.«anatic exploration of a dream described by Moreno (1951) some years ago. It is of interest in that it is both a nightmare and a repetitive dream occurring regularly for a period of 15 to 16 years.

The drama occurred during an intensive one-week residential psychodrama training workshop. The protagonist, during a discussion of psychodramatic production of dreams, had indicated her desire to explore a repetitive dream and the group, about 11 individuals, had agreed to schedule this session for her.

DIRECTOR: Well, Jean's been preparing herself for this evening's session, I suspect. Why don't you join me up here? (*JEAN joins DIRECTOR on stage.*) You have a dream that keeps recurring. (*JEAN nods.*) Recurrent since when? How long have you been having this dream?

JEAN: For the last 15 or 16 years.

DIRECTOR: And what was going on in your life when you began to have this dream?

JEAN: Nothing unusual. Nothing that I know of that could have caused me to have the dream.

DIRECTOR: All right. Now, Jean, I don't want you to tell me the dream, but I do want to know some things about it. Is it a nightmare?

JEAN: Well, actually it starts out sort of pleasant. But it turns into a nightmare before it is over.

DIRECTOR: I see. Now, you've said it is repetitive. Are there any variations to it or is it exactly the same dream every time that you have it?

JEAN: It's exactly the same dream every time. Over and over. Nothing varies in it. Every detail is just the same.

DIRECTOR: And how often do you have the dream?

150

JEAN: Oh, once a week, on the average. It is getting more frequent all the time. And it is growing in proportions. It is on a larger scale, it seems.

DIRECTOR: Once a week. When was the last time you had it? Have you had the dream since you've been here?

JEAN: No. The last time was a week ago Monday night.

DIRECTOR: About eleven days ago. O.K., let's begin right there. Where were you then, when you had this dream?

JEAN: At home. I live in a trailer home.

DIRECTOR: What we want to do to help you warm up to having this dream is to see you as you go to bed. What time is it? What are you doing as you get ready to go to bed? Let's set up the scene for your bedroom.

JEAN: All right. Here is the bed.

DIRECTOR: Single or double?

JEAN: Single. And here's the dresser. The room is paneled in wood colors.
(She spontaneously sets the scene describing the furniture and decor of tht room.)
And over here is a desk and chair.

DIRECTOR: Oh, this room also serves as your office?

JEAN: It is where I live. (Continues to describe the room and set it up.)

DIRECTOR: O.K. Lets see how you prepare for bed.

JEAN: I lock the front door. Turn out the lights. (She is pantomiming as she describes herself.) Come into my bedroom. Turn on the lights.

DIRECTOR: It's all right. Just show us. (JEAN continues to pantomime taking off her clothes, going into the bathroom and taking a shower. Then she puts on a

151

nightgown and goes to bed. She is lying on her side.) Is this the position you go to sleep in?

JEAN: Yes. *(DIRECTOR finds her a pillow, which she puts under her head.)* I keep a night light on. I sleep on top of the covers.

DIRECTOR: Fine. Now, let's soliloquize, Jean. Let us hear the thoughts that may be going through your mind as you lie here going to sleep.

JEAN: I'm *so* tired. It seems that the more I do, the more I have to do. There are so many things that I want to get done.

DIRECTOR: And your feelings. Would you soliloquize your feelings?

JEAN: I'm too tired to think about work. I'll try to think about something nice instead.

DIRECTOR: What do you think about that is nice?

JEAN: Last weekend was the best in nine months with Holly and Heather. We really had fun and just enjoyed ourselves.

DIRECTOR: The best weekend in*nine months. And it is with thoughts about the weekend that you go to sleep?

JEAN: Yes.

DIRECTOR: All right. Relax. Take a deep breath. As you let it out, let the tension in your body go with it *(Pause)* Again. And let yourself go to sleep. Feel yourself drifting off to sleep. *(After a long pause to let JEAN relax.)* How long after you have gone to sleep does the dream begin? Is it early in the night, or later?

JEAN: Right before I wake up in the morning.

DIRECTOR: All right. Go ahead and sleep through the night *(Another long pause.)* It is getting toward tune to wake. What time is it?

JEAN: Around five.

DIRECTOR: Let the dream begin, Jean. Just have the dream again. Watch it in your mind's eye. Let me know when it is over. (*Director has placed a hand on JEAN'S arm. He watches her closely, noting that she trembles during the dream, perspires, and her foot moves slightly. She opens her eyes and has a troubled look on her face.*)

DIRECTOR: Does the dream wake you?

JEAN: Yes.

DIRECTOR: All right, Jean. Now we want to see the dream in action. Let's go to the first scene. Is there more than one scene?

JEAN: -Well, it changes as we go along.

DIRECTOR: Let's begin at the beginning. Where are you?

JEAN: I'm in a car on a country road.

DIRECTOR: O.K. Let's set that up. (*Chairs are placed to represent the car. Jean sits in the driver's seat, holds on to "steering wheel".*) What kind of car is this?

JEAN: A big black one.

DIRECTOR: Are you alone? How fast are you going?

JEAN: Yes, I'm alone. I'm going along pretty fast. I'm anxious to get there.

DIRECTOR: Where are you going?

JEAN: Home, it is Christmas.

DIRECTOR: And what time is it?

JEAN: It is night time.

DIRECTOR: *(Has lights turned down until room is rather dark.)* And what is happening?

JEAN: Well, I'm driving down this black-top road. And I know in a minute when I go around this curve that I'll be able to see the house. I know that my kids are there. It's Christmas and I have a lot of packages in the car. It is cold.

DIRECTOR: And now what happens? What do you see?

JEAN: Now I am up to the fence, a white board fence. I can see the house over there. *(Points ahead of herself and to the left.)* I can see the lights, the Christmas tree lights. The kids are still up. I am driving along and I know that pretty soon I'll come to the gate. And I know that the gate will be open for me. They've left it open for me. *(She continues driving, then begins to look perplexed.)*

DIRECTOR: Yes? What happens?

JEAN: There's just fence here. No gate. I should have gotten to the gate. The fence keeps going on. There is no gate!

DIRECTOR: Lets here your thoughts. *(Doubles)* "Where is the gate! What's going on here? Something is wrong!"

JEAN: *(Nods desperately)* 1 can't get home!

DIRECTOR: Again. Let's hear that again, the way you feel it

JEAN: I CAN'T GET HOME!!

DIRECTOR: What happens next? Have you passed the house?

JEAN: No, the house kind of stays in the same place. I keep driving looking for the gate. *(Director has group become the fence posts moving past tht car.)*

DIRECTOR: What do you feel? Is there more to the dream? What do you want to do?

JEAN: I want to put on the break *(pushes with her foot)* but the brake doesn't work. I take my foot off the accelerator, but the car doesn't slow down. The door won't open! *(Becomes increasingly agitated.)*

DIRECTOR: Brakes don't work. Door won't open. What are you going to do?

JEAN: I'm helpless. I take my hands off the wheel. The car is driving itself. The house is receding. A hill is coming up. I am going to go up the hill. And when I go over the hill, I know that there will be nothing there. *(She is very upset.)*

DIRECTOR: O.K. Jean. Let's go up the hill now. *(He tips her chair backwards.)* Up the hill. *(Pause.)* Now we come to the top *(pauses)* and now we go over. *(Waits a moment.)* What is it like? Are you suspended in space? Are you falling?

JEAN: I don't know. I'm just there. I am suspended.

DIRECTOR: O.K. Show us what it is like. *(Jean gets out of chair and curls up on the floor.)* What is the feeling of this?

JEAN: I don't know what this place is. I don't know how I got here. I don't know why I'm here. *(Pause)* And now I start falling.

DIRECTOR: Let yourself feel that. Let yourself fall. Are you scared again. *(Jean nods.)* Do you feel like screaming? Do you want to scream? Go ahead. *(He puts a hand supportively on her head.)*

JEAN: .*(Screaming)* NO! NO! NO! NO! NO! NO! NO!

DIRECTOR: *(Doubling for the protagonist)* "NO! I don't want. . . ."

JEAN: *(Hesitates)* No, I don't want to die.

DIRECTOR: *(After a period of silence.)* Is this the end of the dream? Do you awaken now?

155

JEAN: *(Nods)*

DIRECTOR: Fine. Open your eyes then. Soliloquize.

JEAN: Damn! Will that never stop? Will I ever get used to this?

DIRECTOR: *(After a few moments silence)* Well, Jean I'll tell you what we are going to do. We are going to let you have this dream all over again. Only this time, I want you to change it in any way and at any time that you want. O.K. Get back in the car. We'll begin at exactly the same place with you in the car on the blacktop road. Do you want to change anything? What kind of a car are you in?

JEAN: It is a big black Lincoln. And it is just full of presents.

DIRECTOR: O.K. Now we are coming to the fence. *(Has group form the fence.)* Do you want to change anything?

JEAN: *(As she drives along)* Yes. The gate is there and I turn in. *(The "Fence" makes a gap and JEAN turns in.)* And I drive up to the house. I can see the Christmas tree through the window.

DIRECTOR: How big is the Christmas tree?

JEAN: *(Smiling)* It's really big.

DIRECTOR: I thought so. What happens next? JEAN: Holly and Heather run out. The twins.

DIRECTOR: How old are Holly and Heather? Let's pick some auxiliaries. *(Two members of the group become Holly and Heather.)*

JEAN: They are ten. *(To the girls)* There are presents for all of you in the back of the car.

DIRECTOR: Yeah. Take them in the house and put then under the tree.

JEAN: Then the rest of the family come out. *(Auxiliaries are picked up for the other four children. They come out and there is much greeting and excitement getting presents out of the car and into the house.)* And then Bob comes out.

DIRECTOR: This is the man in your life these days?
JEAN: Yes. *(Another auxiliary is picked for BOB. He and JEAN embrace.)*
Then we all go into the house, into the living room where the tree is. *(All do so.)*

JEAN: *(Suddenly)* I don't like this living room. Now I remember where it is. It is my grandmother's living room.

DIRECTOR: Well, let's change it then! Make it any room you want. *(As JEAN is thinking this over)* O.K. What's next?

JEAN: Well, now what I would really like is to sing some carols and then exchange gifts.

DIRECTOR: Of course! *(He asks auxiliaries to take it from there, using their own spontaneity. They sing carols. Then unwrap packages and generally produce the happy excitement of Christmas time. They follow JEAN's lead as she produces cues for them. Finally, the activity slows down.)* Well, Jean. Now that you have all the kids here, is there anything you would like to say to them?

JEAN: *(Thinks a moment.)* Just that I'm so proud of each one of you.

DIRECTOR: It must be about time for the kids to go to bed, isn't it? Time for JEAN and BOB to have some time together?

JEAN: Yes. *(She shoos the kids off to their rooms.)* *(To BOB)* Christmas is my favorite time.
BOB: Yeah. I think it's my favorite time, too.

JEAN: I'm so glad you are here to share this with me. It means so much to me. *(JEAN and BOB continue to chat cosily. Finally the DIRECTOR speaks.)*

DIRECTOR: Looks to me like it may be time for Jean and Bob to go off to bed too. *(They laugh and the group joins them. The DIRECTOR dismisses the auxiliary and has JEAN re-set the scene to her bedroom in the trailer, where the original dream had taken place. She lies down on the bed.)* I'll bet this dream doesn't wake you up. Go ahead and dream it as long as you want. Then wake up when you are ready.

JEAN: *(Lying alone. After a period of time wakes up.)*

DIRECTOR: The sun's come up. *(Asks for lights to be turned up. Has JEAN join him in center stage.)* Shall we stop here? How do you feel?

JEAN: Relieved. Lighter. Hopeful. I think I also feel more confident. I can't integrate all this right now.

DIRECTOR: No, I don't expect you to. You've done a lot of work. I think we should stop here. We could explore more, but I think you have done enough for this evening. We will save the rest for another time if we need to. My bet is that you won't have this dream again. And if you should, let me know. I have my ideas on what else could be done. But for now, let's let your natural processes take care of things for a while.

Now is the time to share. *(Group members do so.)*

Discussion

Moreno (1959) identifies four stages or phases of psychodramatic dream production. They are:

1. The original dream.

2. Psychodramatic re-enactment of the dream.

3. Psychodramatic extension of the dream.

4. Post-psychodramatic effect on the dream. Let us examine Jean's dream from this structure.

The status nascendi of the original dream is in the mind of the dreamer during his sleep. The production is solitary, unshared by others. The dream may, as in Jean's situation, be repeated over and over again, reminiscent of a broken record. One gets the impression that it has not been finished. This phase has happened before the psychodrama proper begins, of course.

The second phase typically occurs in two parts. First, the protagonist is warmed up to the moment in which the dream was originally dreamed. Moreno writes (Moreno, J. L., 1969, p. 157):

> In a psychodramatic dream presentation we say to the protagonist "Don't tell the dream, but act it out." We don't mean it only in a sort of superficial way. Let the protagonist go to bed. Let him re-enact every detail . . . The protagonist goes first into the role of the sleeper before he can be a dreamer . . . We try to recapitulate the natural process of living, instead of just analyzing in an unrelated way. That's concretization of the situation in which the dream is presented in the here and now.

This process is apparent in Jean's drama as well as in Moreno (1951, 1969). This procedure serves to develop and increase the "warming up" to the psychodramatic enactment of the dream itself, the next step.

The depth of production is enhanced by avoiding a verbal account of the dream. Telling the dream, putting it into words, tends to conserve it. As sociologists have pointed out, an individual's language system is highly colored by the culture in which he learned it. The psychodramatist's goal is to go beneath the language to the "actional level" and foster communication from there. At the same time, the director may need some information about the dream in order to assist the protagonist in staging it effectively. In Jean's drama, the director explicitly tells her what kind of information he wants and asks her *not* to relate her dream. This principle is also observable in the dreams presented by Moreno.

The protagonist is asked to set the scene, select her auxiliaries, and produce the dream in action. Although every psychodrama is a unique event and the choice of particular dramatic techniques must be consistent with the protagonist's needs for expression, there are some general principles which are specifically applicable to dream production. The first of these has to do with the very personal aspect of the dream. All the spontaneity in a dream is one's own. There is no counter-spontaneity as there is in interaction with other individuals. Therefore, although

159

auxiliaries are used quite freely, even to represent apparently inanimate objects, such as the white board fence in Jean's dream, it is extremely important that the protagonist's perception of the dream be followed as minutely as possible. In the enactment of the dream, the auxiliaries do not alter or expand the roles into which they are cast as is often allowed or encouraged in the enactment of an encounter.

The third stage, the extension of the production of the dream, is probably the most crucial and important phase of the entire process. Here, if the principles of spontaneity are appropriately applied, is where the repetitive dream may be laid to rest, the nightmare expurgated, and the "important dream" integrated. On occasion, the direction in which to take the drama is obvious. One simply continues on from the point at which the protagonist indicates that the dream ended when it was originally dreamed. At other times this procedure is awkward or the director may feel that some aspect of the dream should be altered in order to allow the protagonist relief (catharsis) from the experience. Here, the procedure is to start from the beginning with instruction to the protagonist to change anything about his dream that he wishes to change. This is generally a very pleasing task for the protagonist and the director can expect him to take to it with enthusiasm. Sometimes, however, the protagonist may be primarily avoidant towards the dream and his modification of it consists of replacing strong or threatening aspects with bland ones, or he will simply try to eliminate and cut out action which he did not like. The effectiveness of the drama will be enhanced if the director encourages the protagonist to expend at least as much energy in producing the modified dream as he did in producing the original.

During this phase, auxiliaries' counter-spontaneity may be released to a greater extent. This allows the director, if need be, to use this resource in warming the protagonist up to his task. It is still, however, the protagonist's dream.

In Jean's drama, both extension and modification of the original dream occurred. Extension occurred at the point in the drama when the protagonist began going up the final hill into some kind of a void. This is the point at which she always awakened when experiencing this as a night dream. The director took her on, into the void: Here a significant meaning of the dream emerged. It was obvious that the dream reflected some concerns regarding the protagonist's own death.

The director also felt that it would be most helpful for the protagonist (and the group) if she were allowed to re-do the dream in a different manner. Hence he asked her to return to the beginning of the dream and to make any modification in it that she wished. He was relatively confident that she would complete the fantasy in keeping with the original happy direction in which it seemed to have started.

Moreno's fourth phase deals with the effects of having psychodrama-tized a dream. It raises the crucial question which all therapeutic interventions must face: What has been achieved by the procedure?

Because both Jean and the director are members of the same university community and belong to an ongoing psychodrama group, there has been a chance for considerable discussion between them on this question. During the first few weeks following the drama, Jean reported that the dream had not returned. She was sleeping longer and sounder for which she was grateful since she had developed a lung infection and required more rest. She didn't believe, however, that the dream would not return. She would settle, she said, for just a few weeks of freedom from it.

Later, approximately two months after the psychodrama, she reported that she had looked back over the notes that she and one of the group members had made of the session and she found that she did not want to spend any time thinking about the experience. She related that the whole thing seemed kind of distant, and although she knew very well that it had happened, it didn't seem so real. As she was talking, she said, "You know, I think I am beginning to believe that I am through with that dream."

Approximately four months after the drama, the protagonist and the director discussed in detail the changes that had taken place an her life following the drama. She identified four areas of her life in which she had made significant modifications.

The day following the conclusion of the workshop, Jean had walked into her trailer with her daughter who remarked, "This is like living in a terrarium, Mom." Jean was immediately struck by the fact that her house was entirely decorated with browns and earth tones. "It was like living underground," she commented, "like in a tomb." That had seemed all right until then, but it suddenly became distasteful to her and on that very day, she found another place with bright colors and moved.

The second change involved patterns of sleeping. All of her adult life, Jean stated, she never seemed to need more than 6 hours. This changed immediately and she began to sleep 6, 7, even 8 hours. "The time that I slept 11/a hours was just unbelievable," she said. She also found that she could, for the first time since a child, take naps in the daytime. Associated with this was a general decrease of tension.

The third difference that she noted was in driving. She stated that she had not been aware of it, but suddenly realized diat when she was driving, especially in unfamiliar areas and on unfamiliar roads, she was apparently always looking for, always expecting to come upon that stretch of road which she had traveled so many times in her dreams. She stated that she now realized to what lengths she had gone to avoid driving an unknown road at night. Since the drama, she said, her confidence in driving has increased manyfold and she was no longer afraid of going anywhere, any time.

The most important alteration in her behavior was a rapidly developing self-assertiveness which manifested itself especially with respect to her job and to her ex-husband. She found herself getting irritable, more demanding, and "even aggressive" in both situations. For the past year, Jean had invested herself heavily in her new job. She realized, she said, that she had done so largely for her own "therapy." Suddenly she was aware that she was devoting entirely too many hours a week and too much of her energy to the mental health agency for which she worked. In response to some threatened cutbacks in funds, which would have affected her salary, she resigned, re-wrote her job description for two people, and some weeks later accepted one of the new positions.

Jean had left her marriage in some desperation, knowing that there was something unhealthy about the relationship between her husband and herself. She had, however, never quite been able to deal effectively with her ex-husband's expectations that she would some day return to him. She knew that she did not want to, but being unable to rationalize her feelings, she still retained some uncertainty and some feeling that he just might be right Following the drama, she reported that for the first time she had felt real anger toward him. This was sufficiently bothersome to her to explore it further psychodramatically. Following this, she began bringing about some closure to that relationship. She noted that "I had always managed to excuse Jerry because I knew why he acted that way. I could never get angry at him. Now I can. I don't have to go on making excuses for the way he behaves."

The meaning of the dream remained a puzzle for some time. Although it was obvious that the message of the dream had something to do with death, Jean was unsure of what it might be since she had had occasion to reconcile herself to dying some years previously. She related that she expected death to be very much like the void in her dream, and even though she didn't want to die, she did not really fear it particularly. Eventually, she decided that the dream probably reflected her fears of a living death, of not living her life as fully as she possibly could. In light of the changes she made following the drama, this interpretation made a lot of sense.

After four months, Jean had not had a recurrence of the dream.

The dramatic impact of her drama upon Jean's life raises questions about how psychodrama works, as well as the relationship between dreams and daily behavior. Although the marked changes in her behavior are not simply the result of a single psychodrama session but reflect a readiness on her part and probably a long period of preparation, it is also apparent that the session played a major role in catalyzing and unblocking, allowing Jean to put into action desires and needs that had previously been unfulfilled.

Conclusion

Because dreams are so very personal and come from the innermost parts of oneself, the psychodramatization of a dream is frequently a very intense, meaningful experience for both protagonist and group. The basic rule for the director is "handle dreams with great care and gentleness."

The results of dream production tend to be great. Repetitive dreams seldom return. A sense of relief and catharsis accompanies dramatization of a nightmare. A new understanding of the processes underlying dreams is common. But the most important feature of psychodramatic dream production, which distinguishes it from earlier approaches to dreams, is that it gives the individual a chance to experience his dream in a greater reality, and in so-doing, to gain autonomy over his dream patterns (Moreno, Z. T., 1965).

REFERENCES

Moreno, J. 1951. Fragments from the Psychodrama of a Dream. *Group Psychotherapy,,* 3, pp.   344-365.

Moreno, J. 1969. Psychodrama of a Marriage, a Motion Picture. In *Psychodrama,* Vol. Ill, Beacon, N.Y.: Beacon House,

Moreno, Z. 1965 Psychodramatic Rules, Techniques, and Adjunctive Methods. *Group Psychotherapy,,* 18, pp. 73-86.

ROLE REVERSAL WITH GOD

```
WHAT WOULD THERE BE TO CREATE
IF THERE WERE NO GOD TO BE CREATED
IF YOU COULD NOT CREATE ME?
```
-The Words of the Father

John: J. L. Moreno created a theology-philosophy, a sociology, a psychology, and a therapeutic method. The latter, psychodrama, has eventually achieved wide-spread acceptance after initial rejection. But, as Zerka Moreno (3) has pointed out, the theoretical-philosophical foundations of psychodrama have not attained as extensive understanding or acceptance. A joint account of a psychodrama, in which Cindy was the protagonist, Jim a group member, and I the director, is presented here because we believe that it reflects the integrity of Morenean theory and practice, in this case highlighting the relationship between his theology and psychodrama.

Jim: Paul Johnson writes (1, p. 446) that the basic motivation for all of Moreno's work is religious: "The theory of interpersonal relations is born of religion." Moreno's theater, an open stage in the center of a room with audience access from all sides, invites both actors and audience to portray their own dramatic situations and respond to one another without written lines. For Moreno, this process becomes a kind of dramatic religion, a theater to call forth the spontaneous creative self and "to learn with God what it means to be a creator" (p. 447).

I had the feeling that in Cindy's psychodrama we were dealing with an aspect of psychodrama that is very basic, especially during the role reversal in which Cindy portrayed the role of God.

Cindy: My psychodrama with my father took place six months after his death. His death was sudden and completely unexpected. Our relationship had been the source of much agonizing on my part and when he died, I felt very guilty. I also felt a lot of anger because during the last year I felt we were starting to become a little more human with each other. I felt a sense of outrage that we would never have the chance to become really close.

John: It was the third class session and I was following a procedure much like Warner (4) has described as the "Didactic Chair". The class, about 25 people, had all been asked to bring to mind someone with whom they had unfinished business. Then four volunteers had been invited to enter the stage area where empty chair encounters were developed. Cindy, who had told us that her significant other was her father who had died six months ago, was the last protagonist to engage in the encounter. I was somewhat surprised that she (or anybody else) would bring up so tender an issue at this early a meeting of a rather large class, but proceeded by asking her: "What is your notion of where your father is now? What is your conception of what happens after death?"

"That is an interesting question," Cindy replied.

Cindy: For a number of years, I have considered myself a hard-nosed athiest and was really down on religion. But some time in the last year I had softened my position—not really towards belief, but rather to a less rigid non-belief. After my father's death I went back to being pretty negative. I didn't want to go to the funeral because I was afraid I would get mad at all the garbage they would be preaching over him.

Anyway, I always thought that when you died, that was it. There was nothing else. But I had this experience about a month or so after my father's death. I was driving down the road and saw this rainbow. And suddenly I had a strong feeling of my father's presence—like he was in that rainbow. It was a moving experience which had a great effect on me but I didn't fully understand it.

John: We selected an auxiliary to portray Cindy's father and I asked her to go ahead and tell him whatever she wanted.

Cindy: This was the most difficult part of the psychodrama for me. The first thing I said was the hardest thing to say. Finally, after what seemed to be an interminable amount of time, I looked at him and said, "You know, I'm really sorry you died." I immediately turned to John to explain why I had said that. My father had been away for three years working in another state and during the whole time I had never really wanted him to come home. I didn't miss him. I had timed my visits home from school to avoid being there at the same time he was. And so it seemed really weird to me that now I really missed him. Talking about these feelings was really hard for me because I felt ashamed and guilty at not missing him before.

John had me continue to talk to my father, reversing roles several times. We discussed our relationship and several of the misunderstandings we had.

John: As Cindy became more warmed-up, it was apparent that she felt quite angry that her father had died just when it seemed that they were learning to relate meaningfully to each other. I asked her toward whom or what she felt her anger.

Cindy: I remember saying in reference to God: "How can I be this angry toward someone I don't even believe in?"

John: I directed her to go ahead and reverse roles with this "Non-existence" in whom she did not believe. We were in a room which had steep risers, and I asked her to move to the highest position possible.

Jim: A double was selected for Cindy and posed the question to God: "Why did you take my father just when we were learning to relate to each other?"

Cindy: I enjoyed being God. I remember talking to John, but I don't remember exactly what was said. As God, I wasn't particularly concerned with the girl sitting down there. I recall saying that I had to answer to no one but myself. She wasn't really considered in her father's death.

167

Jim: At .this juncture in the psychodrama, the director began a dialogue with God which had high import for me. I had expected a stance of "plea bargaining" for Cindy, but instead the director merely sought confirmation that the death of Cindy's father was independent of any thoughts, wishes, or actions by Cindy. In response to the question of "Why...?", God in all munificence elaborated that there was no requirement to explain His/Her actions to anyone, no matter how capricious they appeared at the moment.

Cindy: She ("Cindy" portrayed by an auxiliary ego) asked me in the role of God a few questions, but, as I said, I didn't feel I had to justify my actions to her or anyone.

I reversed roles and talked to God. He wasn't really responsive but I don't remember being unsatisfied with the answers He gave me. Then He was dismissed from the scene. I talked a little more with my father at this point, even joked with him. And then I felt satisfied to let him go be a rainbow again.

Jim: Part of the potency of the psychodrama seems to lie in the ability to allow the protagonist to go beyond the reenacting of situations in his or her everyday life. In the case of Cindy, she could reverse roles and become the personification of the Creator—in whom she even expressed non-belief! When Cindy was playing God, I sensed a radiance of power emanating from her. Her pronouncements as God were potent, majestic, matter-of-fact, and self-sufficient. No justifications were offered.

The preface to *The Words of the Father* (Moreno, 19) emphasizes that there is no communication with God better than direct communication. Hence, in order to exist meaningfully we must find the path of creativity and let it lead us into direct communication and identity with the Creator. In this bond, God is represented as the Absolute Creator of the world, much the way Cindy represented him. The preface goes on to note that "His creativity seems so powerful and so all-persuasive as to leave nothing

168

for any other agency to accomplish" (p. xi). Further envisioned is a universal inter-dependency between the Creator of the universe and all the beings who fill it, hence a fellowship with each living thing. This aspect I saw reflected in the Director's straightforward approach to Cindy as God.

After participating in Cindy's drama, I was struck by a passage in *The Words of the Father* (p. 107) which seemed uniquely appropriate:

HAVE YOU LOST YOUR FATHER?

HAVE YOU LOST YOUR MOTHER?

HAVE YOU LOST YOUR WIFE?

HAVE YOU LOST YOUR ONLY BROTHER?

HAVE YOU LOST YOUR YOUNGEST SISTER?

COME TO ME WITH QUICKENING PACE AND REST YOUR HEAD
    UPON MY LAP.

TAKE YOUR PLACE AT MY RIGHT HAND OR AT MY LEFT.

I AM YOUR MOST PERFECT FATHER.

I AM YOUR MOST PERFECT MOTHER.

I AM YOUR MOST PERFECT WIFE.

I AM YOUR MOST PERFECT BROTHER.

I AM YOUR MOST PERFECT SISTER.

John: J. L. Moreno indicates in many places that all his contributions were inspired by his original concerns with religion and the concept of God. Having conceptualized God as the spontaneity-creativity in the universe, he suggests that all the phenomena for which man has resorted to the postulation of a God (or gods) could be better understood as manifestations of spontaneity-creativity, a force or energy or principle which is scattered throughout the universe and found in each individual member of the human race. He thus advanced the concept of God from the distant, demanding Father God of the Hebrews, the "He-God"; and the not-so-distant, but certainly separate, Christian God of Love, the

"Thou-God"; to the "I-God." God is no longer a separate entity or force "out there" (or in "heaven") but a force which occurs partly in each "me," each self. To know God, to be in touch with God, is to be in touch with that part of myself, with my own spontaneity-creativity.

It is fitting that psychodrama provides a direct and concrete means of accomplishing this feat getting in touch with one's "I-God." Not only does this occur in the dramatic fashion which Cindy's psycho-drama illustrates; every psychodrama, 'sociodrama, or role training session has the effect of developing, re-newing, regenerating spontaneity in all those who take part in it.

It is not unusual for God to appear in a psychodrama. Every protagonist has his own perception of God and produces Him in his own way. However, it is remarkable how often this procedure of reversing roles with God produces some kind of sense of reconciliation or relief, even for people who have no belief in the God they have learned about through the church. Frequently a catharsis of anger is involved. The experience of enacting God is usually a profound one and has a positive effect upon the protagonist. Jim: I think that Johnson (1, p. 447) puts it well when he says that in psychodrama we "learn with God what it means to be a creator."

Cindy: My reaction to the experience was positive and powerful. I was absolutely elated for almost a week afterward. I think it was a good way to tie things up for me. Although I have had a few bad times since then I've been able to handle them better. I was also able to write this letter which had been struggling to come out for months:
Dear Dad,
in the months since you died I have wanted to talk to you
more than I ever did
in the years that you lived.

there are things I need to say.

we never talked.
we argued a lot
and on rare occasions we chatted
but
we never talked.
I wish we had.
sometimes I tell people
we never touched
but that's not true.
we horsed around a lot
we could pound each other on the back
but we never hugged each other.

I wish we had.
God, I wish we had.

the last time I saw you
you were lying in a hospital bed
dying
and I tried to tell you then that
I love you
and I tried to thank you
because I never did.
and I wish I had.

I am sorry
that I could never treat you like a person.
in your life

I hated and resented you
I'm sorry
if it seems that only
in your death
have I realized
that I also loved and respected you.

you taught me a lot
not through the very few lectures you gave
but by
the person you were and
the life you lived.

you taught me
to be my own person
and even though we often disagreed
on what I did with my life
I could always feel
the respect you sometimes grudgingly gave me
for being my own person.

you taught me
a lot about responsibility
and even though we sometimes disagreed
about where your responsibilities were
I hope you can belatedly feel
the respect I can now give you
for fulfilling the responsibilities you chose.

you also taught me

how to skip like Dorothy
in the Wizard of OZ.
and that was fun.

I am proud
to be your daughter
and, you know, when I think about it
you were probably proud
to be my father.
which is nice.

I still have guilt feelings
but I guess
we can't do it over again now,
can we?
your life is over.
so what I'm left with is
my life.
and I guess the best thing I can do
for both of us
is to live my life
like I feel it should be lived.
you'll probably roll over in your grave many times
but I think you'll still be proud.

sometimes I just sit and think what
if it weren't for you
I would have never been.
it really flips me out.
Thank you

for the life and the love you gave me.

Love, Cindy

by the way,
I was right about Nixon.

REFERENCES

Johnson, Paul E., "Interpersonal Psychology of Religion: Moreno and Buber", in Greenberg, Ira. *Psychodrama: Theory and Therapy.* New York: Behavioral Publications, 1974.

Moreno, J. 1941. *The Words of the Father.* Beacon, N.Y.: Beacon House,

Moreno, Z. 1970 Moreneans, The Heretics of Yesterday Are the Orthodoxy of Today. *Group Psychotherapy,*, Vol XXII, pp. 1-6.

Warner, D. 1970. The Didactic Auxiliary Chair. *Group Psychotherapy,* Vol. XXII, pp. 31-34.

PSYCHODRAMA IN THE TREATMENT OF
INCEST AND OTHER KINDS OF SEXUAL ABUSE

Introduction

In the past fifteen years' mental health professionals of all disciplines have seen a staggering increase in the number of clients whose problems or symptoms turn out to be associated with or traceable to sexual molestation, often incestuous, during childhood and early adolescence. The tremendous escalation in the number of reported cases of incest and child sexual abuse has resulted in the establishment of a variety of programs involving courts, child protective agencies, welfare agencies, and therapeutic programs which have been developed with the intention of encouraging the discovery of incestuous occurrences, prosecution and/or rehabilitation of the perpetrators, and treatment of the victims. The increase in the apparent incidence of childhood sexual abuse has been so dramatic that initially professionals who were confronted with it wondered if this was a new social phenomenon or if incest and other forms of sexual molestation have been happening all along and simply not being identified. There is plenty of evidence, including the ever growing number of adult women acknowledging this kind of experience, for the current consensus that sexual victimization of children is not new, has been happening longer than there have been mental health professionals, and is responsible for generations of misery, emotional disequilibrium, and psychopathology.

It is interesting in this respect to recall that Sigmund Freud informed the Vienna Society for Psychiatry and Neurology around the turn of the century that in every one of the 18 cases of hysteria which he had treated with his new psycho-analytical method, he had found evidence of premature sexual experience in early childhood years, sometimes assault by a stranger, but more often molestation by a care taking adult, or inducement into sexuality by a sibling. He got a cold and rejecting reception from his medical colleagues for suggesting that sexual abuse was the cause of hysteria and later backed off from this position, arriving at the conclusion that most of the childhood seduction reported by his patients had never happened. In his *Three Essays on the Theory of Sexuality*, published in 1905, he publicly refuted his early view, substituting for it instead the notion that it was the wish-fulfilling fantasies of his patients which lay at the bottom of neurosis. Thus, he came to the astounding conclusions that not only do patients invent tales of seduction, but that it is guilt from repressed, early childhood desires that causes neurosis, and that premature sexual experience, even if it has happened, is essentially irrelevant!

Even though he recanted his original hypothesis (that the emotional disturbances of his patients were the result of childhood sexual abuse), current events suggest that perhaps he was right the first time, and that there is nothing new with respect to the frequency of incest and childhood sexual abuse. One current writer (Masson, 1984) believes that Freud was never so really sure about the correctness of abandoning the seduction theory, and that this fact has been suppressed by psycho analytic historians and biographers. Masson postulates that Freud may have bowed to the pressure of his medical colleagues and public response to his early formulations.

Another psychoanalyst, Freud's erstwhile close friend, Sandor Ferenczi (1932) read a paper to the International Psycho-Analytic Congress in which he expressed his belief that the psychoanalytic profession was not paying sufficient attention to the part that childhood sexual experiences played in the lives of their patients. In this paper, "Confusion of Tongues between Adults and the Child," Ferenczi describes with tremendous sensitivity, understanding, and deep compassion for the victims, the

176

devastation which premature sexual awakening of the child by an adult can wreak. This essay and Freud's original Aetiology have been reprinted by Masson as appendices to his *The Assault on Truth*, making them readily available to contemporary therapists. Both papers are well worth reading. It will be obvious that Freud and Ferenczi have indeed encountered and dealt with the effects of early childhood sexual abuse and understand well the insidious effects which this experience can have in the life of an individual.

Unfortunately, by the time Ferenczi delivered his paper, the psychoanalytic profession was deeply entrenched in the dogma that the unconscious fantasy was of far more importance in the patient's life than was the concrete experience, and Ferenczi met the same stony reaction from his psychoanalytic colleagues as Freud encountered from the Vienna Society for Psychiatry and Neurology 46 years previously.

Psychoanalytic doctrine, in this respect, seems to have served both psychotherapists and their patients ill by encouraging therapists to doubt their clients' stories of childhood molestation and to ascribe them to wish fulfilling fantasy instead of acknowledging them as valid experiences which have very likely happened. Survivors of sexual abuse are thus left to question their beliefs about what has happened to them, and this is especially undermining of the self-esteem and well-being of those who have repressed much of their memories anyway, and are left even more uncertain and anxious about what has happened to them.

Current estimates of the occurrence of child hood sexual abuse seem to be on the order of 22 to 38 percent, and there is enough consistency to suggest that 22% serves as a solid if conservative estimate.

Considerable attention has been given the subject of incest and extra-familial sexual abuse of children and adolescents in a number of publications in the past decade. *Betrayal of Innocence* by Susan Forward and Craig Buck (1978) provides a comprehensive discussion of the varying dynamics typical of all the possible combinations of incest, illustrated with case studies. But it is the title, which echoes Ferenczi's observations about the effects of incest upon the victim, that so succinctly captures the essence of the impact of incest that makes it popular reading for survivors. The Kempe's (1984) *The Common Secret* includes a wider spectrum of sexual abuse than does *Betrayal of Innocence*, and is a bit more scholarly. Included is the outline of a Model Criminal Diversion Program, which provides a therapeutic alternative to legal punishment, and an extensive list of educational materials, pamphlets, audiovisual programs, and books. Some of them are listed in the bibliography below.

My personal experience with an increase in the frequency with which childhood sexual molestation, often incest, became a frequent issue in the psychodrama groups that I was then leading, began in 1976. I was conducting a three-day demonstration psychodrama workshop, attended by approximately 45 people, of whom a majority were women. Fairly early in the workshop the issue arose when a participant psychodramatically confronted a step-father who had introduced her to sex when she was about 12 years old. I was surprised at the number of group members who shared personal experiences of early sexual trauma with her. Her forthrightness apparently encouraged other survivors and rather quickly we had more dramas dealing with various aspects of childhood sexual abuse. My recollection is that more than 20 women in that group acknowledged some form of sexual abuse either as children or adults. Since that time, the number of students and clients reporting childhood sexual molestation and abuse has rapidly increased. I have participated in programs which were specifically established for victims of incest and other sexual abuse, and I have also worked with a number of individuals, mostly but not exclusively women, for whom the trauma of sexual victimization is the event which has brought the individual into a therapeutic process. In addition I have had a number of trainees and/or

178

clients who have come to the conclusion during training or therapy that they had been molested during childhood but had "forgotten" that fact. I have also worked with a number of incest perpetrators.

The experience has been an educational and a consciousness-raising one for me. We live in a culture which is still heavily male oriented and male dominated, despite some inroads achieved by the feminist movement during the past twenty-five years. Most males in our culture (including myself) grow up with quite erroneous ideas about male-female relationships and sexuality. This includes practically no awareness of the terrible impact that premature sexual awakening can have upon a child and that child's emotional future, even if it is not accompanied with emotional and physical abusiveness—which it often is. It is not uncommon to find incestuous fathers who believe that they are initiating their daughters into the joy of sex and fully expect them to enjoy it. This is not simply a rationalization but their actual belief. Childhood sexual abuse will continue to be both a social problem and a source of intense pain with all its emotional and behavioral problems for millions of individuals until these attitudes have been eradicated from the culture and replaced with the widespread, accurate knowledge of the destructiveness of early childhood abuse. While we need to learn, as therapists, how to best help those who have experienced this trauma to live productive and satisfying lives, we need to expend considerably more effort than we have so far in preventive endeavors.

The Impact of Childhood Sexual Abuse

The experience of incest or molestation of children can cause emotional disequilibrium which is disruptive to the survivor for the rest of the survivor's life. Childhood sexual trauma reaches life-threatening proportions and many victims and survivors are suicidal and self-mutilating even years after the abuse. There is no way to count the number of sexually abused individuals who have eventually succeeded in committing suicide. Incest is commonly found in the history of individuals with several severe forms of psycho- pathology, including borderline and multiple personality disorders, both of which are extremely resistant to change and difficult to treat. Commonly, victims and survivors of child hood sexual abuse experience severe anxiety, guilt feelings and lowered self-esteem, sexual problems ranging from frigidity to promiscuity, severe problems in social adjustment and in establishing intimate relationships, and all manner of somatization symptoms including migraine headaches. Eating disorders, which can also be life threatening, including anorexia, bulimia, and obesity, are also frequently associated with early sexual molestation. Many survivors experience problems with alcohol and drug abuse.

At the same time, it has been estimated that a number of molested children, perhaps 50%, do not show pathological residuals. One wonders, after being confronted with the impact that sexual abuse has had on some individuals, how others could come through the experience essentially unscathed. Since childhood sexual experiences range from sexual touching to vaginal, oral or anal intercourse, and from gentle seduction to violent and repeated rape, possibly combined with a threat of death, one might expect a correlation between the degree of violence and the degree of emotional disturbance. It is not quite that simple, however, and individuals who have experienced non-violent, short termed sexual attention may demonstrate significant residuals. For example, a young professional woman presented with a severe obsessive-compulsive disorder which was specifically related to her father's fondling her breasts several times when she was 13 years old. Thus there must be many other factors besides violence or extent of the abuse which play a role in determining the relationship between the sexual experience and its impact upon the survivor.

Some of the obvious factors include the general healthiness of the family, the personality of the victim, and all of the events and conditions which generally influence the development of emotional disturbance and psychopathology.

Father-daughter incest almost always occurs in a severely dysfunctional family, one in which the father typically has adopted the philosophy that home and family are the father's property, and that he is free to do with the members of the family anything that he desires. This is a common viewpoint in many sub-cultures in this country, and, although he knows it is not socially accept able, the father sees nothing intrinsically wrong in his incestuous behavior or is able to rationalize that it is not wrong. The mother of the victim either joins the father in this philosophy, or suppresses her objections (or else they don't stay married) and hence, is unable to come to the assistance of the victimized daughter. Quite frequently as clients, these women report telling the mother of the father's behavior and with the result that they are not believed and may even be scolded for telling or punished for "making up" such a story. Others state that they were afraid to inform for fear of being punished, a situation which suggests that the family is one in which the child is generally wrong and typically gets punished for complaining about any discomfort. This is common in families in which the father is a tyrant. Sons are often favored while the daughters are treated as sub-human.

The personality of the victim, even at an early age, can make a significant difference in the impact of sexual abuse. One young woman, dealing with severe effects from the experience of repeated and severe abuse over a period of two years between the ages of 7 and 9, became aware that she had experienced even earlier sexual initiation at the hands of an uncle and her grandfather. The psychodramatic exploration of this led her to the conclusion that the major impact of these experiences was to make her more vulnerable to later victimization. She did not feel that any of the tremendous difficulties which she experienced as a result of the later abuse would have resulted from the incest itself. After much consideration, she made the decision not to confront the grandfather, who was still alive, even though the popular literature on the subject advocates this step.

Clinically, one can consider both victim and survivor of childhood sexual abuse from the perspective of post traumatic stress syndrome (Janoff-Bulman and Frieze, 1983). The diagnostic criteria for post-traumatic stress disorder include :

1. A recognizable stressful event which is "generally outside the range of usual human experience,"
2. Re-experiencing the trauma via intrusive recollections or recurrent dreams, or by acting as if the event was actually happening in response to an environmental or cognitive stimulus;
3. Numbing of responsiveness demonstrable by diminished interest in significant activities, feelings of detachment from others, or constricted affect;
4. Other symptoms including exaggerated startle response, sleep disturbance, guilt, memory impairment or trouble concentrating, avoidance of activities which trigger recollection of the event, and intensification of symptoms when subject is exposed to events which symbolize the traumatic event.

Treatment Issues

Survivors of childhood sexual abuse seldom enter therapy to deal with that issue. For many, memory of the event is repressed. For others, whatever their current symptoms and problems are, the client does not connect them with the abuse. As a matter of fact, not enough attention is paid by many professionals to the possible connection between presenting problems and a history of early sexual abuse. Common presenting complaints involve difficulties in maintenance of intimate relationships, such as marital problems, depression following the loss of a relationship, etc. Suicidal attempts following some kind of disappointment with regard to a relationship are quite frequently the problems with which clients sexually abused as children come to therapists or mental health facilities. In dealing with clients who have been victims of childhood sexual abuse, a therapist may have to work with repression and dissociation, chronic low self-esteem associated with intense and long- term feelings of shame and guilt, intense anger and fear which has been chronically suppressed, and the multitude of ways in which individuals have learned to cope, over the years, with the anger, fears, shame, and guilt which the abuse has occasioned.

Repression, as Freud originally described it, involves an energetic forgetting of an experience or event. Although almost everybody has the experience of selectively forgetting psychologically painful or embarrassing experiences, these generally can be ascribed to suppression rather than repression. That is, the individual recognizes that he/she makes some effort not to think about this or that event, to push memory of the event from the mind when one is somehow reminded of it. Suppressed events are more likely to be recall able when we are reminded of them or decide, in the course of therapy, for example, that we really should or want to deal with them.

Repressed memories do not come back nearly so easily. Often the lifting of repression comes in the form of nightmares or waking "flashbacks." It can take a lot of work and therapeutic intervention, as well as time to recover truly repressed memories. Even when events are recalled, they may appear dreamlike for some time afterwards, and even after successful therapy, the individual may report that there is an unrealness about the actual events. The only examples of repression that meet Freud's description of this mechanism that I have encountered have been memories of abuse, physical or sexual, although I suspect that it can occur in other kinds of severely painful experiences.

Many survivors of childhood sexual abuse have repressed their memories of having been abused. Sometimes it is a therapist who first begins to suspect that sexual abuse is a factor in the problems that the individual is experiencing. In groups, the survivor who is repressing memory of abuse may become extremely anxious when another member of the group is dealing with the issue of abuse, and not be able to understand why. When this happens, the group therapist will want to consider the possibility that the individual has been victimized and should work to discover the cause of the group member's anxiety, whether it is due to abuse or some other factor.

Even when the survivor has not repressed memory of the molestation, dissociation is frequently encountered. This becomes a problem in the therapeutic situation because, when the very painful emotions are stirred up by recalling and talking about the abuse, the client may have a tendency to go into a fugue or fugue-like state which may be quite scary to therapist, client, and other group members. One must also be concerned about the client's capacity for functioning if she/he cannot be brought back to a more normal state of consciousness.

Secrecy and guilt are benchmarks of sexual molestation. Perhaps the most insidious and ubiquitous effect of abuse of children, whether sexual, physical, or emotional, is that the victim almost always blames him/herself for the experience.

Thus the most common result of abuse is a feeling of guilt or shame and a chronic sense of low self-esteem on the part of the survivor of the experience coupled with extreme difficulty in talking about what has happened to them. They tend to be quite afraid of being ridiculed, rejected, ostracized and abandoned if the secret of their abuse is discovered. These individuals sometimes seem prepared to accept the responsibility and blame for almost any misadventure which subsequently comes their way and are easy targets for those other individuals who see themselves as persecuted by the world and are looking for someone to blame for their discomforts and unease. This inability to stand up for oneself leaves the individual open to continuing victimization and being taken advantage of throughout the survivor's life. The sense of shame and guilt also make it extremely difficult for most people to talk about the events from which the sense of guilt originated.

A number of hypotheses have been offered to explain this apparently paradoxical effect, that victims seem to blame themselves for their victimization. These range from the feminist position that females are socialized to accept the blame for any discomfort they feel, to suggestions that victim guilt serves an adaptational function. Complex psychical mechanisms (introjection, identification with the aggressor, etc.) have been formulated to account for this effect. It may be a simple matter of the child believing that if something terrible has happened to him/her, it is because he/she has done something to deserve it. A common message which children receive is that punishment is received when they having done something wrong. Thus, it is a simple matter for a child who perceives herself being punished (hurt) to conclude that he/she deserves the punishment. It may be this widespread tendency to believe in "distributive justice," that is, that the world is generally fair, that makes victims so vulnerable to feeling responsible for, hence guilty of, their own victimization.

Some writers (e.g. Lamb, 1986) have suggested that guilt feelings perform a positive function and give the victim of abuse a sense of being in control. One writer (Janoff-Bulman, 1979) distinguishes between behavioral guilt and characterological guilt. The former is guilt over behavior which may have contributed to the probability of abuse and the latter involves a sense of feeling that one is an inferior, weak, or other wise unworthy individual compared to others.

Perpetrators often reinforce the victim's guilt. Victims are commonly informed by the perpetrator any number of statements ranging from "This is our secret and you mustn't tell anyone" to "If you tell your mother, she will be real mad at you and punish you severely" to "You are a bad per son. This is your punishment and if you ever tell anybody, I'll punish you even more." Threats to kill the victim or the victim's family are often made by extra-familial molesters.

It can take considerable effort and time to correct this notion of guilt and responsibility at the emotional level. While many adults who were molested as children have already figured out intellectually that they are not responsible for being molested, the feeling that there is something intrinsically wrong with them persists. This discrepancy between the logical and the emotional can be the source of much discomfort and mala-daptive behavior. A group and shared experiences of others can be of help in ameliorating this dynamic.

Guilt will be all the harder to deal with in those survivors who have derived some kind of satisfaction from the molestation. This happens most often when the survivor has been seduced with some tenderness, usually from an individual of whom she is fond prior to molestation and who is otherwise severely deprived of any kind of attention and love. The situation, of course, is one in which the child is willing to undergo the unwelcome sexual advances in return for some positive attention. A molested daughter may develop a sense of being "important" to the father, and it is certainly not unknown for incest, even when it is not welcomed by the child, to result in the child feeling competitive with the mother. The therapist must be willing to explore all these issues honestly without being prematurely supportive to the client. Such support not only is ignored but possibly results in eroding the trust between therapist and client.

186

Associated with guilt is low self-esteem, a problem which can get in the way not only of daily living, but can also impede the therapeutic process itself, when, in a group the group member hesitates to express what is bothering him/her because other group members "probably aren't interested". Thus it becomes crucial to help the survivor move to a position in which the responsibility for the act of abuse is assigned to the per perpetrator. And this must be accomplished not only at the intellectual level but at the emotional level as well. Often one has the sense that with respect to the molestation itself, that the client is functioning as emotional child and an intellectually mature adult. It is tempting, and usually futile, to try to reason the individual into feeling angry with the perpetrator rather than feeling guilty.

Finally, the therapist needs to keep in mind that the experience of sexual abuse which the client has suffered has become a major building block in the rest of the client's life. It has entered into the individual's social development and choice of companions, into career choices, and into all other major life choices and especially those directly involving sex, such as marriage and family. The decision to confront the abuse and deal with the emotional aftermath of it can have great impact upon the client's whole life scope. Dealing with the abuse may change the client's relationships with family of origin, spouse, children, and other areas of life.

Psychodramatic Treatment of Sexual Abuse

The following discussion assumes that the reader has at least a rudimentary understanding of the psychodramatic method. For those who don't the example described beginning with the paragraph after next should give the naive reader some sense of the psychodramatic process. Childhood sexual abuse can generally be looked upon as post traumatic stress syndrome. The essence of psychodramatic treatment of it involves the psychodramatic re-enactment of the traumatic event or events. The following description is an example of how psychodrama was used in this regard.

In a three-day psychodrama training workshop in which childhood sexual abuse appeared early and had become a common theme of the psychodramatic work which was being done, a participant brought another workshop member, Alma, to me during a break on the third morning. She was a young woman in her mid-twenties and told me, with the support of her companion, that she had been experiencing increasing discomfort during the course of the workshop and had a splitting headache as we talked. She related the discomfort to the theme of sexual abuse and told me that she had been raped at age 12. She didn't think that she could deal with it in the group. After we talked a few minutes, she decided that she would consider it, and in the afternoon session presented herself as a protagonist.

A few minutes of interview revealed that the rape was a gang rape at the hands of three older boys, one of whom was the best friend of her 16 year old brother. This boy's younger sister was 12 and was Alma's best friend, and she, too, was present and helped restrain Alma during the rape. I asked her what aspect of the experience bothered her the most, and she replied that it was her guilt, her feeling that she could have avoided the rape and didn't. I suggested that we re-enact the events leading up to the rape and see if we could figure out where she might have made a mistake.

The individuals involved in the rape, besides her friend, Jane, and her brother's friend, Paul, were two older boys, Kurt and Tim, perhaps 17 or 18. Her father, a man with strong religious convictions who was also the neighborhood social worker, had gone out of his way to befriend these two who came from dysfunctional families on the block.

It was a Sunday evening early in autumn, and Alma's mother and father had taken Alma, her brother, Jane, Paul, Kurt and Tim to youth fellow ship and they had just returned home. Tom, Alma's brother had to go in and finish up some homework. Someone suggested, "Let's go the school and play," and the other five headed out to a school about two blocks away.

This was re-enacted as a psychodramatic scene. The next scene saw them approaching the gate to the school yard which was fenced in. As this occurred, the protagonist broke out of role to announce to the director, "This is where I made a mistake."

"Then let's see it," I replied. As they went through the gate, Alma reversed roles with one of the older boys. "When we have new kids around here, they have to get initiated," "he" said.

In her own role, she soliloquized, "I wonder what he means by that. That kinda scares me." She broke out of role again and announced, "That is when I should have turned around and run home. Instead, I started running inside the playground. That's when they caught me and pinned me to the ground."

"Why don't you turn around and run home?" I asked.

She thought a minute. "I didn't know what they were going to do. I didn't think that they would really hurt me."

"Right," I replied. "Who are you here with? Your best friend, Tom's best friend, and two kids your dad has tried to help out. You might have had reason to think that they might do something unpleasant to 'initiate' you, but you would have had to be a pretty paranoid little girl to believe that they were going to rape you, don't you think?"

She considered and replayed the moment over in her head. Then she nodded her head and agreed.

We continued with the drama. In role reversal, she demonstrated how the boys had caught her and threw her to the ground. Now she was scared. Someone across the street came out on the porch and she screamed to him for help. He went back into his house. The boys pinned her to the ground with her friend Jane helping. She struggled but they were too many and too strong for her to free herself. One of the auxiliaries placed himself over her in the position of the perpetrator. There was no need, of course, to carry the psychodramatization any further, and she told us how each of the boys had taken a turn at raping her.

At this point, I ended the scene and suggested that we do something which might be helpful. She volunteered that she had already learned something. "I had always thought that maybe if I had struggled harder, I could have gotten away from them," she said. "Now I know that I couldn't have. I tried as hard as I could and I couldn't get away."

"Yes," I added. "And when you were 12, they had even more advantage over you than your auxiliaries do today." She agreed.

Next I asked Alma to pick another member of the group to play her 12 year old self. When she had done so, the auxiliaries replayed the school-yard scene. When they had the auxiliary "Alma" restrained, I invited her to enter the scene and do something about it. Without a moment's hesitation, she rushed over to the grouping and one by one pulled the tormentors away from her auxiliary self. They struggled hard as but under these more fair conditions she overpowered them individually and scattered them to the far corners of the room. In effect, she rescued herself.

Then she embraced the auxiliary playing her 12 year old self. I encouraged her. "This young girl is going to do some terrible things to herself as a result of this. She thinks that she is responsible for it. What do you think? What do you want to tell her."

She responded by talking to her younger self. "This is not your fault," she insisted. "It is theirs. They are sick. They had no right to do this to you. You did not deserve this. You did not ask for it!" And so forth. Then I reversed roles and she listened as a 12 year old to her present-day self. And nodded in agreement.

190

The scene ended. I thought that perhaps the drama was drawing close to an end also. "What happened after that?" I asked.

"I was hurt and I was bleeding. The other kids ran away and I went home." She paused. "But I couldn't tell my mother."

"Why not?"

"Because I just knew that it would upset her too much."

"Has she ever found out?" I asked.

"Oh, yeah. About a two weeks later," she answered. "I went to see the school nurse and asked her about pregnancy. She asked if I had any reason to believe that I was pregnant, so I told her what had happened. She told me to go home right that minute and tell my mother."

"And were you pregnant?" I asked.

"Yes."

We dramatized the scene of telling mother. Alma was right. Mother became hysterical and called father who was also so upset that he came home from his office without hanging up the phone! However, with all of their emotional turmoil, there was nobody left to be calm except Alma, and she found herself in the position of being supportive to the parents instead of other wise. Much of the rest of the story was related with out being re-enacted. The pregnancy was con firmed. The parents and her doctor insisted on an abortion, and Alma expressed feelings of guilt for agreeing when she didn't really feel right about doing that. The three boys were charged with rape, and one day a police cruiser had come to the school and taken Alma to the station where a lie detector test was administered. She was not told the results and worried about whether the test indicated that she had lied or not. The two older boys were convicted and sentenced and made threats to find her and kill her when they got out of prison.

In the process of relating all this, Alma realized that she was angry with God. So we let her have an encounter with God. How, she wondered, could God let such a terrible experience happen to her. And then, how could He let her get pregnant and have an abortion. In the middle of this, she said, "And the lesson that night in youth fellow ship was 'turn the other cheek!'" Now in a rage, she demanded of God, "Just what did you want me to do?"

I asked her to reverse roles with God and an auxiliary took her role. "How do you feel toward Alma, and how does it make you feel to see what she had been through?" I asked. "What do you think of her anger toward you, and what do you have to say for yourself?"

She paused for a moment. Then, "I love her very much and it hurts me deeply to see how she has been hurt. I am not angry that she is angry with me."

"What do you feel like doing?" I demanded.

Tears came into her eyes. "I would like to comfort her."

"Then do it," I demanded and she embraced the auxiliary playing herself and sobbed deeply. "I am so sorry," she repeated over and over.

"Can you explain why this has happened to her," I asked. "I think that she has interpreted it as a test of her faith and feels like she may have failed it."

"Oh, no," she replied, "I didn't cause this to happen to her. I have given people a choice, and they choose to do wrong to her. I don't cause bad things to happen to people. They do it to each other." She continued to explain and to comfort from this role.

In role reversal, the scene was repeated and after some give and take, Alma was ready to bring it to a close. She looked relieved, centered, and relaxed. The drama itself was over.

At this point, as should always happen, the other members of the group shared with the protagonist their experiences and feelings which connected them to the protagonist. Sharing was profound. The group was large, at least 45 people, both men and women. Every single person was deeply touched by Alma's story. Some, who had themselves been sexually abused, shared those experiences. Some talked about the experiences of someone close to them. Others shared their fears and fantasies of being so victimized.

It is not often, however, that one attempts to cover so much ground in a single psychodrama, and, indeed, re-enacting the actual event of abuse in the client's first session is more rare than common in my experience..More often than not, one must pay heed to the psychodramatic rule which calls for proceeding from "periphery to the central." In this context it calls for dealing with the more current effects of the abuse before tackling the most crucial one of the abuse itself. Because pf the power of psychodrama, the psychodramatic therapist must be careful to insure that the trauma is treated and not repeated.

With Alma it was possible to accomplish so much because she was a highly functioning individual, in spite of the traumatic sexual experience she had undergone, and had the emotional and intellectual maturity to confront herself at quite a profound level.

Most clients with a history of sexual abuse do not seek treatment for the residuals of that experience. Rather they present a variety of other problems when they seek therapeutic help, whether or not they have repressed their memories of the sexual abuse. Depression, marital problems, family problems, general problems in getting along with people, are often the presenting complaints. It is only during the therapeutic exploration of their current problems that unresolved emotional residuals from the abuse become apparent and are made a major focus of therapy.

A frequent beginning point, when the survivor recalls the abuse, is a psychodramatic confrontation of the perpetrator by the survivor, managed by having one of the other group members play the role of the perpetrator. Appropriately warmed up to this action by the director, this situation is both safe enough and real enough to generate strong feelings in the protagonist, typically some combination of fear, anxiety, and anger. The expression of the protagonist's feelings (ventilation or abreaction) is encouraged, and the protagonist may be asked to reverse roles with the perpetrator, at least once.

The director will attempt to update the protagonist's feelings. That is, the feelings that occur in this situation will usually represent some degree of regression. They are the same feelings that the protagonist experienced as a child at the time of the molestation. At that time, the protagonist may well have had reason to be afraid of the perpetrator, or was afraid because of his threats. In the present, the perpetrator might well have more to fear from the protagonist than vice-versa. The protagonist is more apt to see that from the role-reversal.

Fear often stands in the way of the protagonist expressing the anger which is usually appropriate to the situation, and until the fear has been expressed and dealt with, it will be difficult or impossible for the protagonist to express the ethical anger which is felt toward the perpetrator. Ethical anger is that which arises from injustice done unto one, as opposed to anger which may be defensive and used to hide other emotions. The expression of ethical anger is the protagonist's first step in correcting the damage which the abuse has caused in her life and toward resolving feelings of guilt and low self esteem.

For some protagonists, psychodramatic confrontation of the perpetrator is too threatening to begin with, and some other starting point must be found. For the survivor who has maintained secrecy about the abuse, even from her closest relationships, this may be a psychodramatic rehearsal for informing someone close to her of that fact that she has been abused. From the psychodramatic perspective, this is called role training and the purpose is to prepare the protagonist to carry through with the rehearsed behavior in real life. This is based upon the premise that it is

194

helpful to the survivor to inform someone close to them about the abuse and to be able to talk with them about it.

Many incest survivors appear to be more angry with the mother than with the perpetrator. Their position is that the mother should have been able to protect them from the father (or brother, uncle, etc) who molested them. With these protagonists, the initial session may be a confrontation of the mother.

Those individuals who have repressed the memory of the abuse present different problems. Sometimes childhood sexual abuse becomes an issue because of their reactions to other protagonists dealing with it, and sometimes they are referred by other therapists who have recognized symptoms of abuse. The director may begin with these group members by psychodramatizing the situations in which the individual has symptoms. This might be re-enacting a scene in which the protagonist experiences intense anger at the father without knowing why, or perhaps has gotten quite angry and overreacted to a sexual overture from someone in her daily life. Dreams, usually night mares, and quasi-hallucinatory experiences often are a tip off to repressed memories of trauma. The dramatization of these experiences may stimulate the recall of repressed material.

In one such psychodrama, the protagonist created a scene in which she had been raped at what seemed to her to be about age 3. During the enactment of the scene, which began as Playback and then moved into psychodrama proper, she experienced severe physical pain. Some pain continued during the week following, and although she felt sure that it was associated with the psychodramatic event, she consulted her gynecologist who informed her that he found objective findings of irritation. It looked, he told her, as if she were recovering from surgery, even though he knew that she had not had surgery.

Although she (and I) do not yet accept the scene we created as a faithful reproduction of what may have happened to her, we are both convinced that something must have happened to her at about that age. Another survivor commonly awoke at night with a sense that there was something in the bedroom with her which was scary to her and to which she reacted by closing her eyes so that she wouldn't see it. When we dramatized this experience, it developed into somebody who came and got in bed with her and did something sexual to her. She eventually recalled vivid and precise details of repeated visits from her father in the middle of the night over a period of 5 or 6 years.

When it comes to the psychodramatic re-enactment of the events of abuse themselves, the psychodramatic therapist needs to proceed with considerable caution. Especially with protagonists who are prone to dissociation, the director must be alert to catch the first signs of the protagonist "leaving" and make efforts to stop the dissociating. This often means backing away from the stimulus for the painful emotional response and approaching it in a safer, more gradual manner. Sometimes it means utilizing therapeutic dissociation. This can be done by playing a threatening scene playback style. The protagonist, in this technique, is allowed to sit off the stage and an auxiliary is chosen to take the protagonist's role in the drama. Then, after the protagonist has expressed some of the suppressed emotionality, the scene may be played with the protagonist included in the action.

With individuals who have repressed memories of abuse, the therapist can expect that from time to time, the survivor will suddenly go back to doubting that the abuse ever happened, wondering if they have made all this up. This can occur even many months after the survivor has recovered rather detailed memories of abuse. One psychodrama group member still had such doubts a year after she first began suspecting that her father had forced intercourse upon her at about age 7. She had recalled being molested, though never raped, by an older brother much earlier in her therapy. Even after she began remembering details of the fathers abuse of her, she would become quite upset and wondered if she was just making this up in her mind. At one point, she had become rather self-destructive and was experiencing a severe abdominal pain which we had already established as associated with rape by her father. During a drama in which she was protagonist, she affirmed for herself again that the father's abuse had actually happened: "I'm NOT just making it up!" and the pain went away as well as did the self-destructive behavior. For awhile.

One of the problems the psychodrama director faces is how to stage the abuse. Needles-to-say, one does not commit the abusive sexual act again. On the one hand, the purpose of the psychodrama is to bring about a redintegration of the original emotional experience. On the other hand, the protagonist is an individual who is there because she has suffered one of the most damaging emotional experiences imaginable. Directors find all manner of creative ways in which to concretize the molestation in non-harmful ways. One must discover from the protagonist what is necessary and what is too much. For example, stroking her bare arm has been quite sufficient for protagonists who were fondled fathers and brothers. Another wanted to re-experience her molestation more concretely and was able to do so by asking a female auxiliary to lie on top of her. This raises the issue of asking a group member to be an auxiliary, to take and play the role of the perpetrator. This, in itself, can cause some emotional distress for the chosen group member who may wonder if the protagonist sees some capacity for such behavior in him/her. The director will want to be sensitive to any hesitation in a group member asked to play a role with such negative over tones, and may have to be sure that taking the role will not be counter-therapeutic to that group member.

Sometimes it is not the exact touch or act of the perpetrator that the protagonist needs to re-experience but the helplessness or the terror, and these can be concretized by dramatizing the trap, immobilization, or threat which the protagonist describes.

Courtois (1988) describes a philosophy for the treatment of incest survivors which comprises three principles. The first involves recognition of the reality of childhood sexual abuse and the fact that it has disruptive effects upon the victim. The second principle rejects the psychodynamic and traditional medical models of psychopathology in favor of traumatic stress and feminist approaches, supplemented by family system models to conceptualize the abuse and its aftereffects. And the third principle recognizes the imperative of tailoring the therapy to the individual who is being treated. All of these principles are easily assimilated into the psychodramatic approach, if not intrinsic to it.

Courtois also discusses counter-transference reactions which, of course, are counterproductive to the therapeutic process. These include the stirring up of feelings of dread and horror in the therapist when confronted with the client's report of incest, and some of the stories that survivors have to tell involve almost unbelievable events of abuse. Freudian theory had, until recent years, conditioned most therapists to deny and avoid the reality of incest and childhood sexual abuse in favor of the Oedipal theory of the sexual wishes of the child. The therapist who cannot get beyond this perspective can only be iatrogenic to the victim or survivor of childhood sexual abuse.

Other emotional responses which the therapist must be able to control include feelings of shame, pity, and disgust, or guilt and rage at the perpetrator. The first three can lead the therapist to distancing by mobilizing his/her own defensiveness against them. Guilt on the part of the therapist who has lead a happy childhood, or on the part of the male therapist through identification with the gender which is most frequently abusive can lead the therapist to overextend themselves to the client, even encouraging dependent behavior, while rage can interfere through the therapist attempting to express the client's rage for her or by pushing the client into anger before the client is ready to accept it and deal with it.

Other counter-transference traps include perceiving the client as all victim, fragile and helpless and in need of someone to take care of her, or as all survivor, a heroine with superior resources and coping abilities, in which case the "child within" can be neglected in the therapy. The client's experience can be trivialized by the therapist who emphasizes that all women are victims, minimizing the individuality of the client who is seeking relief from her specific experience of abuse.

Both client and therapist can engage in language muting, avoiding the use of explicit and descriptive terms in favor of less emotionally laden ones. "I am a victim of sexual abuse" rather than "My father sexually molested me" or "I am an incest victim." Therapists are vulnerable to contact victimization, that is, to developing sympathetic symptoms of post traumatic stress syndrome, and must guard against an inquisitiveness into the details about the incest based on personal dynamics, privileged voyeurism.

Courtois suggests that incest therapy may be more complicated for male therapists due to socialization and power issues. There is also some data which suggest that incest survivors are more frequently sexually abused in therapy than other female clients. On the other hand, she points out that with male therapists "who are secure in their sexuality and who address and work through power issues...[and] can then identify with the victim" the survivor "has the opportunity to experience a caring relationship with a man with whom appropriate sexual boundaries are maintained and to learn to distinguish sex from affection." (p 241.)

The female therapist, on the other hand, is susceptible to over-identification with the victim and, of course, to negative transference when the victim deals with her anger at the mother who didn't protect her. There is considerable debate about the therapist who has been a victim of sexual abuse. On the positive side, this individual certainly will be able to empathize with the incest clients. On the other hand, this therapist must have worked through his/her own issues thorough ly or risk counter-transference reactions which can be damaging to the client.

Therapy with Incest Perpetrators

My experience in working with perpetrators is more limited than it is with regard to the treatment of victims. It includes primarily perpetrators who have accepted treatment in a diversionary program as an alternative to imprisonment and this means it is a population rather carefully selected with a good prognosis. They are all fathers. Although these men had engaged in a wide variety of sexual activities with daughters, none of them had been involved in the more violent acts such as forcible rape and other such brutalities.

Upon entering treatment, the majority are angry, embarrassed at being found out, and enraged at having what they perceive as their private family life invaded by punitive law enforcement officials and nosey welfare people. The therapist who is going to work with this population must be fully prepared and skilled in dealing with resentment and resentful clients. "Resistant" is far too mellow a word to describe the average perpetrator's initial attitude and a small percentage of them would have stayed in treatment without the threat of prison hanging over their heads.

The first task is usually to communicate to them that they have done something wrong. Even with a felony conviction, the seriousness of which his parole officer will periodically remind him, the perpetrator typically holds a traditional male chauvinist position, and believes sincerely that a man's wife and children are his property, his home is his castle, and he is constitutionally entitled to do what he wants in whatever way he chooses.

200

Experience with victims of incest is quite helpful to the therapist who is in a position to reinterpret the impact of his behavior on his daughter to the perpetrator.

On the other hand significant numbers of perpetrators will enter the program in deep shame, readily acknowledging the enormity of their behavior. They are sometimes depressed and suicidal. The initial steps here involve helping the client realize that he can redeem himself and even be forgiven by those whom he has violated.

The second task generally is getting the perpetrator to accept full responsibility for is actions. There is a strong tendency to blame someone else, the victim, the wife, hard knocks, alcohol. Success in preventing a recurrence is heavily dependent upon the perpetrator acknowledging that he, and he all by himself could have prevented the incest from happening. While other factors may play a preventive role, he is the most important one.

This procedure involves identifying the "set up," or in psychodramatic terms, the "warm-up" to molestation. In other words, the perpetrator is asked to examine the pattern of his incestuous behavior. For example, he obviously must arrange to be alone with the victim. How does he arrange this? Perhaps the simplest preventive device is the avoidance of this situation.

A common phenomenon with these men is alexithymia, an inability to read their own feelings. This does not necessarily mean that they are unaware of feelings, although some are, but that even when they are obviously emotionally stirred up, they cannot identify or label emotions. Many are either depressed or in a rage state, and their feelings are essentially suppressed and unavailable to them.

Helpful therapeutic maneuvers involve the exploration of their feelings and identifying their "warming up process," how they prepare themselves for action. Typically, they are not very reflective people and it is difficult for them to examine themselves and their lives. Hence this is often an indicated action. It is even more difficult to get most of these men to discuss the actual molestation than it is with the victim, and so of course, I insist that this be done and that it be done in detail. This is often perceived by the individual as punitive on my part, that I am trying to embarrass him and his fellow group members. But there is good reason to do it. It resolves some of the denial and it can be used to point out to him what he often denies, that there is a period of preparation in which, if he will be in touch with himself, he still has the control necessary to abort an approaching molestation. The extent to which therapy is successful or working can often be perceived in the increasing ease with which he will describe the pattern of his molestation and accept the responsibility for it. This indicates that he is working with the therapist rather than resisting therapy.

Beyond convincing the perpetrator that he has indeed committed a grave injustice upon his child, and that he must take full responsibility for his behavior and develop his own program for preventing a recurrence, therapy can focus upon reestablishing a healthy relationship with the daughter. The first issue is usually the rebuilding of trust between them. In this respect, the perpetrator's guilt can sometimes be an obstacle. An example is one father whose guilt was expressed in a hang-dog demeanor in the presence of his daughter. Portrayed psychodramatically, he was able in role- reversal to identify that to her, it was a consistent plea to be forgiven, and subtle and indirect way of placing an additional responsibility upon her shoulders. The position that we seek is something like: "I have done a serious injustice to my daughter. This does not mean that I am a bad person, even though I have done a bad thing. My responsibility is to prevent a recurrence, and to atone by becoming the best father to her that I can." It is helpful with these individuals to help them reach a position of forgiving themselves.

It maybe surprising to learn that many of these men, once they have gotten "hooked" into therapy, and have gotten over their initial defensiveness, appear to have good fatherly qualities. When therapy is successful they may appear genuinely caring and loving toward their daughters, and are genuinely repentant and contrite about what they have done to them. The re-establishment of responsible trust between father and daughter is, of course, probably the most healing experience that the incest victim can have.

Prevention

Although the therapeutic collective, and hopefully, society at large, has been somewhat overwhelmed by the discovery of so much childhood sexual abuse, the more horrible thought concerns all those millions of individuals whose stories have not been told, who have not had a chance to repair the damage which has been done them.

While all the programs which have been developed to encourage disclosure of abuse and to provide treatment to the victims and their families are to be lauded, it is far more urgent to consider the problem of prevention. It is not enough to treat victims, even if that could be done with far greater success than we are now capable of achieving.

There was a time in mental health circles when it was fashionable to speak of prevention. With the incursion of private sector business practices into the medical and mental health areas, with its emphasis upon "bottom line" or profit, the notion of prevention seems to have become obsolete. Although there may be much to gain from prevention programs, there seems to be little profit in it. In this area, however, those of us who have worked with aftermath of childhood sexual abuse know that it is unthinkable not to put great effort into prevention.

Bringing about changes in attitudes which have had negative, disastrous effects for females in our culture has been the agenda of the Feminist Movement for many years. There is no doubt that it has been the effectiveness of this movement, especially in the last twenty-five years, that has led to the emergence of the facts about incest and childhood sexual abuse in our culture. It is certainly time for all therapists, for all individuals who take upon themselves some responsibility for the society in which they live, to join in this endeavor. It is probably not necessary to remind psycho- dramatists in this regard, that one of Moreno's basic tenets is that we are co-creators of society, that we do have a responsibility to discover and exercise our creativity in this regard.

REFERENCES

Allen, C. (198Q)*Daddy's Girl: A Memoir.* New York: Simon & Schuster Armstrong, L. (1978) *Kiss Daddy   Tonight: A   Speak-out on Incest.* New York: Hawthorn

Courtois, C. (1988) *Healing the Incest Wound.* New York: W. W. Norton & Company, Inc.

Forward, S. and Buck, C. (1978) *Betrayal of innocence: Incest and it's Devastation.* New York: Penguin   Books.

Ferenczi, S. (1932) Die Leidenschaften der Erwachsenen und deren Einfluss auf Charater und Sexualentwick lung der Kinder. Read before The 12th International Psycho-Analytic Congress, Wiesbaden, September, 1932. Published in *International Zeitscrift fur Psychoanalyse,* 19 (1933), pp 5-15, under the title, Sprachverwirrung zwischen den Erwachsenen und dem Kind (Die Sprache der Zartlichkeit und der Leidenschaft); re-translated and published as Confusion of Tongues between Adults and the Child-The Language of Tenderness and of Passion, in Masson, J. (1984) *The Assault on Truth.* New York: Farrar, Straus and Giroux. Appendix C.

Freud, S. (1886) The aetiology of hysteria.Read before the Society for Psychiatry and Neurology, Vienna, April 21, and published in *The Standard Edition of the Complete Psychological Works of Sigmund Freud.*

Janoff-Bullman, R. (1979) Characterological versus behavioral self-blame: Inquiries into depression and rape. */. Person. & Soc. PsychoL* 37(10) pp 1798-1809

Lamb, S. (1985) Treating sexually abused children: issues of blame and responsibility. *Amer. J. Orthopsychiat.* 56(2)   pp 303- 307.

Masson, J. (1984) *The Assault on Truth.* New York:Farrar, Straus and Giroux
Rev. 11/26/90

A PSYCHODRAMATIC PERSPECTIVE ON RAGE IN GROUP PSYCHOTHERAPY

Rage and it's unbridled expression are issues of grave concern in our society. As these problems manifest themselves at alarming rates in families and public settings as a whole, clinicians are faced with new challenges in treating the rageful patient.-Program description, Session 313, Annual Meeting A.G.P.A., 1994

This paper is a re-working of the author's participation in the Multidimensional Survey Track, *Dealing with Rage in Group Psychotherapy,* of the 1994 Annual Meeting of A.G.P.A, which included taking part in a panel discussion, *Viewing Rage in Group Psychotherapy from Various Perspectives* and conducting a workshop, *Dealing with Rage in Psychodramatic Group Therapy.*

Rage, of course, is expressed through violence and aggression, and the word calls forth images of primitive madness, of a madman on the rampage. Dictionary entries include: "angry fury, violent anger;" "a fit of violent anger;" "violence of feeling, desire, or appetite;" "a violent desire passion;" and as a verb: "to act or speak with fury; show or feel violent anger; fulminate." The word rage is derived from the Latin *rabies,* meaning madness.

Rage, considered as a condition or state of an individual which can lead to violent acting out, is an uncomfortable phenomenon to encounter under almost any circumstance and it can be especially anxiety provoking to the group psychotherapist who has to be concerned for the safety of him/herself, the safety of the other group members, as well as the safety of the rageful client and others upon whom he might express that rage. The purpose of this paper is to offer an analysis of rage, based on the author's experience of resolving intense anger in psychodramatic group therapy, and to describe the psychodramatic process for dealing with rage.

What Is Rage?

Rage can be conceptualized as a state or condition involving a number of emotions, each of them seeking expression in actions which are, to some extent, incompatible. A combination of at least three factors is involved, and a fourth element is usually, if not universally, present. The first component is anger, often a combination of several related emotions, anger, contempt, and rejection. Rage develops when an individual is confronted with a situation in which something or someone is provoking intense anger. In other words, the person experiencing rage is in a situation which is highly frustrating since anger is a natural emotional response to a barrier between the subject and an important goal. The function of anger is to energize the individual in preparation for removing the obstacle standing between the individual and satisfaction. One can say that the function of anger, as is the function of all the emotions (De Rivera, 1977), is to transform the relationship between the subject, who experiences the emotion, and the object of the emotion.

The second component of rage is the presence of a constraint against the expression of that anger, usually fear. The one in a rage, the subject, is afraid of the consequences of expressing his/her anger, afraid of punishment or retaliation; or he/she is afraid of loss of control, loss of self-respect, or of damaging one's image of oneself. Something of this sort restrains the expression of the anger that is felt. It is the suppression of expression which results in the intensification of the anger to the point that it is commensurate with the condition called rage. As long as the fear is sufficiently strong to inhibit expression of anger, and as long as the subject remains in the frustrating, anger-provoking situation, the individual tends to maintain a state of rage. It is when inhibition fails that rage will be expressed in violence. On the other hand, if the frustrating circumstances are mitigated, the anger, and hence the rage, tends to decrease or even disappear.

The third component of rage is psychic pain, sorrow, or grief, and its presence is often not apparent to either the subject or to observers until after the angry feelings have been expressed. Recognizing this element helps account for those incidents in which an estranged spouse or rejected lover kills the erstwhile loved one in anger, and then kills oneself out of grief.

The final factor common to the condition of rage is the perception on the part of the subject that he/she has been betrayed by someone or something which the subject trusted implicitly. Betrayal is consistent with the other elements described above, and helps account for the intensity of the feelings of anger at the object, the conflict or ambivalence about expressing it, as well as the hurt which is often masked by the anger. Rage, then, may be considered as a combination of three basic negative affects, anger, fear, and emotional pain, and is a response to frustration at the least and perhaps to a perception of betrayal. Considered from this perspective, an increase in violence is not surprising in a society which appears to becoming increasingly more complex and difficult to live in comfortably for increasing numbers of people. The shift of control of material resources in our society to fewer and fewer individuals, and the sizeable number of citizens who are homeless, hungry, and unemployed are conditions which might be expected to result in an increase in collective frustration.

This conceptualization of rage is a "bare bones" outline, and the vagaries of emotional functioning allow for an infinite variety in the manifestations of rage. There are two emotional phenomena so commonly associated with rage that they bear mentioning at this point. The first is the fact that when an emotion such as anger or fear is aroused but not expressed, it seems to have the effect of sensitizing the individual so that the affect is evoked with less provocation. Thus a frequently frustrated individual may "have a chip on their shoulder" or may be described as chronically angry and seem to be seeking situations which provoke anger. Generalization, in which the emotional response to an object in a situation becomes attached to other objects associated with it, is another phenomenon common to the state of rage. This means that the expression of violence may spread from the individual perceived as the source of frustration to other individuals connected with the situation as happens in horrifying massacres of innocent victims.

Psychodramatic Approach to Rage

Rage emerges in group therapy in several configurations. A group member may become angry (enraged) at the group leader or another group member. A group member may undergo an experience of intense frustration, such as being fired, being deserted by a spouse, losing a court decision which seems unfair, or the like, and bring the emotional aftermath into a group session. Some group members are in therapy because they have been aggressive or violent. However, perhaps the most common source of rage, encountered in group therapy, stems from the recognition and realization during the course of therapy that the individual has suffered childhood abuse of one kind or another, usually from a parent or trusted caretaker.

The unique feature that distinguishes psychodrama from other modalities of group psychotherapy is its employment of dramatic enactment, its emphasis upon action and therapeutic acting out. In working with rage, the psychodramatic process often follows a scenario which can be schematically outlined something like this:

1. The protagonist (subject of the psychodrama) enacts a recent life event which has provoked rage (intense anger which has either not been expressed or has been inadequately expressed). Appropriately done, the protagonist's anger will be re-energized. The director invites the

protagonist to express that anger in the psychodramatic situation with an "auxiliary ego" as a substitute for the target of the anger. The protagonist may or may not be able to express anger effectively.

2. If the protagonist is unable to express the anger, the cause is explored. This commonly leads to recall of earlier childhood experiences in which the protagonist was overwhelmingly intimidated by a parental or other adult figure. One of these events may be dramatized. Again, there will be suppressed anger at this individual. The director facilitates the expression of this anger.

3. Once this anger is expressed, the protagonist will very likely be able to effectively confront the more recent object of rage and express the anger experienced in the current situation.

4. Following this, the protagonist may experience hurt, sorrow, or grief, and the expression of this affect is facilitated.

5. With the reduction in intensity of the emotions involved, the protagonist is usually ready to explore the situation from a new perspective. New perceptions of the significant other, less encumbered by projection and enhanced through role reversal, or re-evaluations of the self may emerge and give a different meaning to the situation. The intrusion of emotional reactions from past events into current events may be identified and sorted out.

Psychodrama provides certain advantages over non-action group methods in dealing with rage. The use of the auxiliary ego allows the emotions to be focused and the re-enactment serves well to reactivate the emotions associated with the events in question.

The method also provides for the acting out of intense anger in a safe manner. Anger is expressed in a number of ways. Shouting and screaming at the offending person, hitting, kicking, pounding, pummeling, striking, biting, ripping to shreds, throwing something—all are expressions which protagonists have used at one time or another. The protagonist often has a definite sense of the action which he/she wishes to engage in, which is appropriate for the situation. Shouting and screaming, swearing, and epithets provide few problems other than dealing with group members who may be hypersensitive to such actions, often because of their history. Other forms of expression require some forethought and preparation. A well-equipped psychodrama theater will contain a mattress, cardboard boxes, something fairly heavy to throw against a wall, and a wall substantial enough to withstand that treatment, and so forth. When the protagonist feels the desire to kick, a heavy cushion or pillow, or a mattress on the stage can serve as an "auxiliary body" for the kicks.

Confronted with a protagonist who is inhibited in the expression of anger, the director may first try encouragement and may even ask the protagonist to engage in anger expressing behavior even if the protagonist doesn't "feel" it. When this is ineffective, as in the psychodrama described below, the director may next decide to investigate the source of the inhibition, "resistance" in psychodramatic terminology. Frequently, as was the case here, this leads into earlier experiences which, if they are resolved psychodramatically, unblock the protagonist's resistance to action.

A Psychodrama

In the previously mentioned A.G.P.A. Conference workshop, a demonstration psychodrama was produced which effectively illustrated many of the principles of dealing with rage through psychodrama. Responding to the director's request to recall an event "in which you had an encounter with an absolute, arbitrary authority figure", a volunteer protagonist related her ordeal as a psychology student with a practicum supervisor whom she perceived as denigrating of her, unhelpful, and who had put her into awkward situations.[6]

The director asked her to recall an experience which would exemplify his treatment of her. She choose a time when the supervisor had taken her to a locked ward where he was to interview a newly admitted patient, a

----

[6]The situation hardly meets the level we would call rage but the anger was intense enough to illustrate how intense anger might be dealt with psychodramatically.

rapist. The supervisor placed the student, a small and rather attractive young woman, and the patient is a small interview room, and had then left without either instruction or explanation. She was both afraid of the patient and embarrassed at not knowing what to do or say. This event was psychodramatized with auxiliary egos taking the roles of supervisor and patient.

The enactment successfully reactivated protagonist's anger, and she was invited to express it to the supervisor, as represented by the auxiliary. She said, without much intensity, "Fuck you, and I hope that I never have to see you again." The director encouraged an expression of anger more commensurate with the intensity of her feelings. She made several attempts, but each time she hesitated. She turned to the director and said, "I can't do it. Something holds me back." She illustrated by putting both hands on her chest just below her throat.

Exploring this with the psychodramatic technique of concretization led to recall of being scolded by her father (who is no longer living) at age three. With the technical assistance of the director, she re-enacted an experience in which she is three years old, dancing and singing to a television commercial, feeling very happy. Her father comes into the room and shouts at her to quit singing and dancing and turn off the tv set.

The re-enactment redintigrates feelings of guilt and shame ("I feel like I am bad and that I am doing something wrong") in the role of the young daughter. The scene is re-enacted again by auxiliary egos, group members who take the required roles, and she watches as her adult self. As an adult she is angry with the father. When the director asks her what she would like to do, she responds that she has a black belt in karate and would like to throw him to the ground. She is allowed to do this in a way that is safe for the auxiliary in the father's role. The director asks her to reverse roles with the father. The protagonist, in the role of the father, reports feeling

211

regret at having treated the daughter so harshly and acknowledges that "he" took out frustrations from elsewhere on her.

With the protagonist still in this role, the director asks: "If you could live your life over, would you do anything differently?"

As father, she thinks for a moment. "Yes. I wish that I had played with her rather than scolding her. It would have been better for both of us." The director says, "Let's see how that might have been," and directs a scene in which the three year old daughter and the father skip and sing to the t.v. This is followed by a brief and moving encounter in which both express regrets about what had happened and had not happened between them during his lifetime.

After allowing the protagonist to savor this "surplus reality" encounter, the director now brings the protagonist back to her encounter with the supervisor, again inviting her to express her anger. She responds that rather than just blasting him with her rage, she would like to make him understand how abused she felt by him in the hope that he would be more sensitive to other students in the future. The director agrees to let her try, and she expresses her sense of having been abused by him with considerable forcefulness which seems appropriate to the situation.

"Reverse roles," the director asks. She and the auxiliary playing the supervisor exchange places. The auxiliary, now in her role, repeats her charges to the supervisor. "And how does that make you feel?" the director asks the "supervisor."

"I could care less," is her reply in the role of the supervisor. Roles are reversed again. "Now what?" asks the director.

"It's hopeless," she replies indignantly. "I felt no regret at all in his role. I don't want anything more to do with him."

"Then let's get rid of him," responds the director. He directs the protagonist to push the supervisor across the stage space to symbolize "pushing him out of your life space." She does so very effectively despite the fact that she is smaller and lighter than the male auxiliary who was asked to truly resist.

The action part of the psychodrama is now complete and the director asks the group members to share by relating personal experiences of which they were reminded during the drama. Since most of the 35 people in the room are professional therapists, there are many stories involving insensitive supervisors as well as experiences between fathers and daughters to be related.

Catharsis: Abreactive and Integrative

While there was practically no possibility that this protagonist might resort to violence as a result of her relationship with her practicum supervisor, this example does serve to demonstrate the dynamics of rage and to illustrate the psychodramatic approach to rage.

J. L. Moreno identified catharsis as a major benefit of psychodrama, and this has led to some confusion because he extended considerably the definition of catharsis that Sigmund Freud had introduced into psychiatry in 1896 (Moreno 1953, 1972, 1975; Nolte, 1991). Freud described "Breuer's cathartic method" in *Studies in Hysteria* (Breuer & Freud, 1957). Considering emotion to be a by-product of unexpressed sexual instinct, Freud concluded that the "abreaction" of emotion under hypnosis which occurred in the use of Breuer's method cleansed the psyche and provided a cure for hysteria. His initial enthusiasm for the cathartic method faded when he discovered that the patients inevitably relapsed, and he rejected the use it in favor of free association. Catharsis is controversial in the field of psychotherapy today with some theorists and practitioners advocating its use while others consider it useless. A few believe that it is counter-productive in psychotherapy, leading to an increase in aggressiveness (Nichols & Zax, 1977)

Moreno considered abreaction, the expression of strong, unexpressed emotion, to be only one aspect of catharsis, the other being "catharsis of integration." The aim of catharsis, for Moreno, is to bring about a change in the dynamics which exist for an individual in a specific situation. Since the situation generally involves an experience which has already occurred, the only way that this can be done is to alter the individual's perception of that experience. In the psychodrama, the protagonist is given an opportunity to explore and extend or expand his understanding of various elements of a situation. In role reversal, the protagonist has the possibility of experiencing an event from the position of any significant others who were involved. This includes an opportunity to assess the possible motivations and perceptions of these individuals, including perceptions of themselves from the role of the significant other. The psychodramatic situation allows the individual to identify information which may have registered below the level of awareness during the experience. In the process, the protagonist may come to a different understanding of the meaning or significance of the event in his/her life experience. The protagonist may revise his/her perception of how he/she handled a difficult situation, or may gain new insights into the behavior of significant others such as parents or supervisors. In the psychodrama described above, the protagonist initially experienced feelings of guilt and shame when criticized by the father for singing and dancing. Looking at the situation through her adult eyes, she was able to revise this self-evaluation and identify the father as having acted in an unjust way.

A basic principle of emotional functioning, supported by the work of De Rivera (1976), and repeatedly demonstrated in psychodrama, is that when an aroused emotion is given appropriate and adequate expression, the emotional state of the individual changes. However, in psychodrama and other expressive therapies, therapists occasionally encounter a group member who seems to have no end to their anger, generally at an abuser from the past. This client, as a protagonist, appears to be capable of angry expression indefinitely, with little positive gain. Anger and rage, for these individuals may be serving as a defense against the experience of pain and the feelings of vulnerability which accompany it. The director may be able to assist the protagonist to experience and express the pain directly by returning to the abusive events, although that must be accomplished with considerable care (see Nolle, 1984). Another strategy involves dramatizing positive experiences in the protagonist's life, events in which the protagonist identifies his/her psychological strengths.

The seasoned psychodrama director never loses sight of the fact that the abreaction, the expression of strong emotion, is not a goal in itself. One is not trying to "get an emotion or feeling out" of the protagonist, as if it were a toxin or waste product (Fritz Perls referred to Freud's "excremental theory of emotions",) but rather the director is attempting to facilitate the protagonist in understanding his/her life experiences in a more effective and realistic way. Strong, unexpressed negative feelings tend to perpetuate the existing perceptions of self and others, while their expression can facilitate the discovery of new meaning in past events.

Social Implications

If the analysis of rage presented here, which suggests that rage is a direct consequence of frustration, is correct, it is not surprising that violence is on the increase, considering that we live in a society in which it is becoming more difficult for greater numbers of people to "live the good life" comfortably. It appears that material wealth is being concentrated in the possession of a smaller percentage of the population, and this is probably reflected in sizeable numbers of people who are homeless and/or who go hungry in a country which produces more foodstuff than it can consume. Along with burgeoning technological development has come a society which is much more complicated to live in. To be successful today, one must have considerably more education then was necessary just a couple of generations ago, and one must be familiar with new technological processes of communication, such things as computers, facsimile machines and the like. It is preferable to possess and utilize them. The division between the "haves" and "have nots" increases. Despite a massive legislative program which eliminated legal segregation of the Black race, *de facto* separatism not only has not been eliminated but appears to have increased. For twenty-five years, the response to crime in American society has been increased punishment, a technique which has been amply demonstrated in the laboratory and outside it to suppress the punished behavior without altering the probability of its occurrence. In terms of the current presentation, punishment of lawbreakers by imprisonment can be considered calculated frustration, forcibly suppressed. Since few individuals are incarcerated for life, this means that we are incubating violence for the future in the pretense of "correcting" it. In addition, this culture has supported almost unlimited, unregulated distribution of high-tech instruments for the expression of violence—guns, especially handguns. If social engineers somewhere on earth, sometime in the future wish to devise a society rife with violence, they will have little more to do than to follow the blueprint drawn up by American politicians of the past 25 years.

J. L. Moreno was more than a psychiatrist. He coined the term "sociatry" to define the work which he felt was most important, the treatment of society. He could also be considered the first radical psychotherapist. There was no logic, he contended, in insisting that people adjust to society if it was a sick society. Rather, one should help one's patients change society, make it better, more like it ought to be. He also held that anger and violence began between individuals, and that when it was not resolved at that level, it resonated in their social networks, spreading out to others. Thus dissension between two persons could spread to become dissension between two families, and, as it combined with other angry feelings, dissension between two communities.

Eventually an accumulation of unresolved anger becomes dissension between two nations, and is sometimes expressed through war. To reduce violence in the world he advocated dealing with violence and the factors which cause it at every level of occurrence. Psychodrama is one instrument he devised for improving interpersonal relationships between individuals, couples, families, and small groups. Other methods that he originated, such as sociodrama and sociometry were designed to reduce frustration at the community level and higher. Unfortunately they have so far generated too much anxiety for extensive application.

REFERENCES

Breuer, J. & Freud, S. (1957). *Studies in hysteria.* New York: Basic Books

De Rivera, J. (1977). *A structural theory of the emotions.* Psychological Issues, vol. X,   No. 4, monograph   40

Moreno, J. (1946). *Psychodrama, first volume.* Beacon, N.Y.: Beacon House, Inc.

Moreno, J. (1975). Mental catharsis and the psychodrama, *Group Psychotherapy and Psychodrama,*   28, 5-32

Moreno, J. (1953). *Who shall Survive?* Beacon, N.Y.:Beacon House, inc.

Moreno, Z. (1971). Beyond Aristotle, Brewer and Freud: Moreno's Contribution to the Concept of Catharsis. *Group Psychotherapy and Psychodrama,* 24, 34

Nichols, M. P. & Zax, M. [1977]. Catharsis in psychotherapy. New York: Gardener

Nolte, J. (1992). *Catharsis, from Aristotle to Moreno.* Presented at Annual Meeting of A.S.G.P.P. New York,

Nolte, J. (1984). *Psychodrama in the treatment of victims of sexual abuse.* Presented at the Conference of the Western Region Chapter of A.S.G.P.P. Santa Monica,   January.

217

PROTAGONIST WITHOUT A PROBLEM

When I first became involved with psychodrama in 1959, the J. L. And Zerka were traveling throughout the United States and occasionally to Europe, lecturing and demonstrating the psychodramatic method. The first psychodrama that I saw Zerka Moreno direct was in Kansas City at a major psychiatric institution in that city. Her protagonist was a high ranking staff member and, at his request, they explored his problematic relationship with his father. While I was very impressed with her skill and amazed at the power of the drama, I was also bothered by the degree to which the protagonist, through the skill of the director, had revealed his private feelings and relationships. I knew that I would not have wanted to make public so much information about myself and my life experiences. Jonathan Moreno has since (1994) has discussed the ethical issues involved in the public session.

In the next few years I was present at a number of these public sessions and almost always experienced some ambivalence. On the one hand, protagonists seemed grateful to have had the chance to explore issues and problems that were bothering them. They did not seemed particularly bothered by having revealed so much information about themselves. But I remained a bit uneasy about public or open sessions. None-the-less, when I was given the opportunity, I directed sessions, sometimes at psychodrama conferences, where neither I nor the protagonist knew very much about the people who made up our group.

I also had occasion to demonstrate psychodrama to various groups in the communities in which I lived, largely in the Midwest where people may be quite friendly but are possibly more reserved than those who live on either coast. I was even more cautious about inviting members of groups that I introduced to psychodrama to share their personal problems in public sessions.

I found a way to handle these situations in something that I had heard J. L. Moreno say numerous times: "Every meaningful experience should be experienced at least two times; once when it happens and once on the psychodrama stage." All the elements of psychodrama and all the techniques can be demonstrated by psychodramatizing any meaningful experience. And I have done so many times since.

The warm-up is a directed one and quite simple. "Think of a meaningful experience that you have had sometime in your life. What comes to mind?" Sometimes, when invited to recall and share a meaningful experience, members of the audience will bring up painful experiences. I look for a positive experience and then ask "Are you willing to share that experience in action with this group?" I have found that some of the most profound and moving psychodramas have come from dramatization of meaningful experiences. The birth of a child is frequently mentioned and I particularly like to let a father dramatize this event, knowing that at some point he will be in the role of the mother and gain a unique perspective that he can get in no other way.

Events of a child with a grandparent or a grandparent with a child are also common and put both protagonist and audience in touch with the transcendent love that often exists between those separated by a generation. Other experiences from childhood recall for all present the magic of adventures in which we discover something new and wonderful about the world in which we live.

Protagonist without a Problem psychodramas underscore how much a psychodrama is truly an art form, whether or not it is used for therapeutic purposes. Dramatizing meaningful experiences also makes a good training exercise. Students are often reluctant to accept the director role when a protagonist presents a personal problem. Meaningful experiences relieve some of the perceived responsibility while giving the student director a chance to apply all the tools of psychodrama.

REFERENCE

Moreno, J. D. (1994) Psychodramatic moral philosophy and ethics. In Holms, P., Karp, M., & Watson, M. (Eds) Psychodrama since Moreno. Rutledge: London:

THE SCRIPT WALK

There are psychodramas and there are psychodramatic activities. This paper describes a psychodramatic exercise. I developed it one afternoon while teaching a course on psychodrama at Sangamon State University in Springfield, Illinois. There were about 30 students in the class which met once a week. Most of them were counseling or psychology majors although the class was open to all students. I started class meetings by stimulating a discussion, asking the class about any problems they might be having in their classes, especially with the concepts or techniques that were being presented to them. I would then use that discussion to select a pertinent psychodramatic principle or rule which I could illustrate in psychodramatic action. This in turn, provided an experience for discussion.

On this particular afternoon, near the end of the semester, the class was unusually lethargic and discussion lagged. I realized that action was indicated and, possibly out of irritation with the lack of energy which pervaded the whole group, decided to get everybody into action at the same time. I organized the group into sub-groups of six persons each. Then, borrowing from an old encounter group exercise, the life walk, I gave the following directions:

1. "In you own mind, which parent is most responsible for who you are today?" I then asked each person to indicate their choice by simply saying "mother" or "father" as I went around the entire group.

2. "Now reverse roles and become the parent you have chosen, and become that parent *at the time of your birth.*" This is a complicated direction and it takes some time to make sure that everybody has understood and complied.

3. "Here you are with five other brand new parents. Get acquainted with each other and talk with each other about what it is like to be the father or mother of this new baby. Is it your first child or do you have others? Did you or the baby have any problems? Any concerns? Talk about your marriage. Are you and your husband or wife getting along well? Was this a wanted and planned for baby? How is your financial situation? Here is this little new life in your hands, your responsibility. What are your hopes for this child? What are your concerns."

4. When it became clear that the groups were running out of things to say, I gave the next direction: "stand up and move around. As you do move through time until you child has become three years old." After giving the group some time: "Now return to your group. By some miracle the six of you who met when your baby was a newborn have found yourselves together again. A reunion. Your child is now three and, most likely is mobile, able to walk and run, and is talking. How is he or she doing? Is your child's development going along well? Or are there any problems." After the group has had a chance to discuss these questions: "What about family circumstances? Have there been any major changes? Anybody divorced? New jobs? Lost jobs? Any deaths, grandparents for example?" And finally, after this discussion: "At this point what are your hopes and concerns for your child?"

5. Again I asked the group to walk through time until the child is in the first grade. The participants were asked to discuss the same issues and others that are appropriate to the age; e.g. "Is your child doing well in school?"

By now the structure of the exercise should be clear. Since that first time, I have conducted this exercise many times and with many groups. It is very flexible because the leader can chose how many developmental stages to include, as well as which ones to include. I have used earlier ages than three, reminding the group members of the significant developmental issues. I often choose ten years old, a time when extra-family influences like teachers and friends and friends families have been established; twelve or thirteen, puberty and relating to the opposite sex; early high school when one is beginning to establish some independence from the parents. This exercise can be a one hour exercise or a three hour exercise.

Now back to the first time at Sangamon State. As I was conducting this exercise for the first time I decided that graduation from high school would be the final stage. When I announced "And now it is commencement day for this child of yours," a young woman in the group burst into tears, crying a wailing. I moved to her side and waited until the sobbing subsided enough that she could talk. Between sobs she told us her story. She came from an impoverished family, one which apparently had poor for generations. Her father had been her hero and her supporter and had pushed her to be the first person in the family's memory to graduate from high school. He had lived for the day in which he would watch her graduate. And he had died a month or so before that day. She, of course, had chosen him to reverse roles with for the exercise.

So we reenacted her commencement so that her father could have that wonderful and longed for experience which had been denied him in life.

Since then I have developed a less dramatic ending for the exercise. The sub-groups are meeting the day before commencement. The directions are:

"You child is graduating from high school tomorrow and going out upon the world. You have had your major influence upon your child, guiding him or her, developing his or her character. Think about what

223

you have tried to teach your child about life and how you would like for your son or daughter to live their life. Formulate in your own mind what you think is the most important thing that you hope you have taught your son or daughter. But don't share it with your group yet."

Many people need quite a few minutes to sort this out. After it appears that most are ready, I ask the subgroups to come back together as a single group and place two chairs in the middle. The directions are: "One at a time, sit in one of these chairs; place your child in the empty chair and tell him or her what the most important thing about life is. You might phrase it like this: 'What I hope you will always remember (or do) is....'"

After the first participant makes the parental response, I direct: "Now reverse roles, become yourself and sit in the other chair. Respond any way you wish to your father (mother.)" The question will arise: "respond as I was then or as I am now?" and I allow the participant to decide or to do both. The group member then returns to the group in their own role.

The exercise continues until everybody has done this last part and the whole group are back in their own roles.

After the first rendition of this exercise at Sangamon State, I dropped by the office of my friend and colleague, John Miller, and described the exercise to him. John, who was familiar with Eric Berne's Transactional Analysis as well as psychodrama and the life walk, exclaimed, "You have just invented the script walk!" He pointed out that this exercise focused the participant on many of the experiences which influence what Berne called the life script. I have called it The Script Walk since.

The script walk can accomplish several things. Participants get insights and self understanding. The exercise also tends to warm many people up to issues not completely dealt with and hence makes a good

224

opening exercise for a psychodrama training workshop. The group leader can also evaluate the cohesion of the group by how participants respond to the situation. A cohesive group gets warmed up to the exercise, talks more and takes longer to complete each assignment. A less cohesive group will tend to do the minimum with which it can get away.

In forming subgroups I like to make use of an idea that I learned from Martin Haskell, one of the seasoned psychodramatists when I was new to the method. The leader first forms pairs by asking the group to pair up with "someone you would like to get to know better." Then two or more pairs are put together to form the subgroups. If the leader wants groups of four, he puts two pairs together; groups of six, three pairs, and so forth. Subgroups of six or eight seem to work well with the script walk.

One thing that the leader of this exercise needs to be prepared for is the death of a parent before the group reaches the end of the exercise. I have already described how I handled that problem the first time I ran the exercise. Now I often 227see a hand go up from a participant who has reached a stage in the exercise and the parent has died. I give the participant an option of changing to the other parent and continuing from that role, or letting the participant remain as the spirit of the dead parent.

ROLE TRAINING IN MEDICAL EDUCATION

Introduction

Professional education can be thought of as embracing three kinds of learning. The first comprises the conceptual-theoretical material and facts which we generally call knowledge. With respect to medicine, we immediately think of the basic sciences, especially the biological sciences, and of the clinical studies.

Another kind of learning involves professional skills. In medicine, these range all the way from learning to observe and assess vital signs to the advanced and complex diagnostic activities and therapeutic manipulations of the various medical specialties. Recognition of signs of pathology, the reading of X-rays, performance of blood chemistry tests or urinalysis are some of the former; prescription of appropriate medicines and drugs, setting broken bones, and intricate surgical procedures are but a few of the latter.

Finally, there is the learning of role behaviors. This category includes a complex set of social skills which serve to establish a professional identity in the application of the physician's skill and knowledge when he is face-to-face with his patients. Role behaviors have to do with human relations, ethical values, and attitudes, and what was once referred to as the bedside manner of the physician. By and large, the medical training centers of today do an outstanding job in the transmission of knowledge and in the development of professional skills in their students. However, the role learning of the student physician seems left largely to chance, and he develops and learns his professional role primarily through an unmonitored process of role modeling; that is, through imitating his instructors and preceptors. Thus, he tends to reproduce a unique combination of both their strengths and weaknesses in his own enactment of the physician role.

This need not be so. There exists an established, if little-known, technology of Role Training which offers a systematic and effective approach to the teaching of these aspects of professional training, which sociologists sometimes refer to as the socialization process of a profession.

Application of Role Training in medical pedagogy could have a number of desirable and beneficial results both for the individuals so trained, and for the profession as a whole. Successful role behaviors enhance the effects and results of professional skill, maximizing the effectiveness of a physician at any given level of skill development. Greater patient satisfaction can be expected, as role behaviors include skills in relating more comfortably with the persons in the role of patients. Skill in role behavior also enhances the personal satisfaction that the doctor gains from his experience as a physician. This reduces some of the stress which most physicians experience from the tremendous responsibilities they assume in their profession. Finally, it is likely that greater interpersonal satisfaction between doctor and patient could serve to reduce the likelihood of the malpractice problems which have plagued the medical profession in recent years.

Role Training in a Medical Training Center

The number of ways in which role training might be applied in medical education is limited only by the imagination and creativity of the role trainer and other members of the teaching staff. However, there are some obvious applications, some of which have already been demonstrated in one setting or another. Most of the examples below have come from actual practice of role training or of psychodrama, first cousin to role training. Some are especially relevant to psychiatry where these methods have been most broadly applied, while others illustrate the application of the principles involved in a more general way.

As the medical student prepares for his first actual encounters with patients, he often experiences fears and anxieties which can intrude adversely into his first interactions with patients. Although these fears seem to dissipate rather quickly for the average student as he gains some experience, there is an opportunity at this stage of training to assist the neophyte physician to recognize and be sensitive to the human factors which enter into the doctor-patient relationship, and to use this awareness in constructive ways.

In a role training session devoted to dealing with these uncomfortable feelings, students were asked to explore their internal preparation to seeing a new patient: Their warm up to this situation. With some assistance from the trainer and from some more advanced students who had been invited to join the group, the students identified a number of feelings and attitudes — some positive and some negative. Among these was the excitement of getting to engage in the activities for which they had been in preparation so long, of doing what a doctor does, and the anticipation of positive emotional returns from successfully doing so. After some variations upon this theme, some concerns began to emerge. Among these are fears of embarrassment by being awkward and being perceived as being inadequate or incompetent by the patient. Will the patient recognize that I am just a student? There is also the problem to be dealt with of making an intimate kind of contact with another person's body which medical examination requires and which is sanctioned, but which conflicts with earlier learning about the appropriate ways of dealing with strangers. Other fears focused on the kind of impression the student was making upon his instructor, fears of forgetting something important, of causing pain to the patient, etc.

229

The first pay-off came from exposing one's fears and discovering that it is a rare concern which is not shared to some degree by one's fellow students. There is comfort and almost automatic relief of some degree of anxiety by the discovery of this all-too-obvious, but previously undisclosed fact.

Having recognized, identified, and acknowledged the feelings which are common to the student's part of the professional relationship, the trainer then suggested that the group reverse roles and experience the feelings of patients who are about to be examined by medical students. Drawing upon previous personal experiences in the role of patient, and perhaps upon some imagination as well, the students produced reasonable and believable portrayals of a group of patients, waiting to be examined by medical students. They were encouraged to get in touch with the feelings engendered by being in this role, and to discuss these feelings with the other patients. From this activity, they found that perhaps some of their concerns about how they might be perceived by patients are probably justified and correct. But, as patients, they found, they seemed to be more concerned about the medical problems that they might be having and generally discovered themselves willing to respond in a positive way to someone who was understanding, communicated genuine concern, and who generally seemed to know something about what he was doing.

Now the trainer moved into a third phase in which the group members took turns interacting with each other, one playing the physician and the other the patient. Although there was a certain amount of caricature produced by this activity, comic relief so to speak, there was a lot of serious rehearsing for future actual encounters with patients. They gave each other valid feedback and suggestions as to what seemed most appropriate and effective.

A relatively short period of time spent in activities of this sort can serve to reduce the student's anxieties to reasonable, appropriate levels where it can serve to enhance rather than to interfere with his judgement and performance.

It is probably helpful to every individual who works with other people to experience himself from the role of his patient or client. Few of us have a very accurate perception of the impression we are making upon others or of the impact of our behavior on them until we engage in this basic role reversal. We tend to be much more conscious of the impression we *want* to convey than of the one we <u>are</u> conveying! Video tapes are a popular way to obtain information about our behavior, and as useful as they are in this respect, the role reversal provides additional information.

The importance of experiencing himself and his procedures from the role of his patient became very apparent to a physician who was in a psychodrama training group. He had occasion as the protagonist in a psychodrama to re-enact the performance of a pelvic examination on a woman patient. During this process, he was asked to reverse roles with his patient and thus experienced, psychodramatically, his first personal pelvic exam. He reported that this event gave him an entirely new perspective upon this particular procedure, and even though he had always been able to imagine why women tend to dislike it, the experience of allowing himself to be placed in the position of lying flat on his back, crotch exposed, etc. , being entered with instruments, brought home to him in a much more emphatic way than empathy can the emotional and physical discomfort that this common examination can occasion. He also realized how very important the behavior of the physician can be in minimizing (or maximizing!) these discomforts.

In a role training workshop for surgical residents, an eye surgery resident also received an important lesson by reversing roles with a

231

woman whom he had identified as his most difficult patient in recent days. First, he re-enacted his original encounter with her, a young black woman who had shown up at the eye clinic where he trained. After he had taken her role to show her demeanor and how she had behaved with him, another member of the group took her role. He then demonstrated how he had dealt with her. His approach seemed to communicate a genuine and professional concern with her problem, a chronic eye infection for which she had sought treatment a number of times previously. He explained the problem to her as he saw it, gave her a prescription and careful instructions for daily treatment of the eye. He stressed the need for her taking a great deal of care of the eye and told her of the possible complications of neglect. He appeared to be doing a careful and thorough job of teaching her what she needed to know about her problem.

Her response, however, seemed surly and ungrateful. Her demeanor and the expressions on her face as reproduced by the resident himself in taking her role seemed to say, "A lot of good you are doing me! I might as well not have come. You're not going to make me well!" The resident had the distinct impression that she was not apt to follow his instructions, and tried harder and harder to impress upon her the importance of doing so. He began to sound like a parent lecturing to a disrespectful child. Finally, with some loss of patience, he said emphatically, "If you don't treat your eye like I've told you, you could lose your sight in it!"

Her response was little more than an eloquent shrug which implied, "So what's it to you?"

The re-enactment awakened many of the same feelings which he had felt in the original encounter. He was angry—angry at the patient for her apparent refusal to accept his genuine concern and interest in helping her, and angry at what felt like rejection of his professional opinion and advice. He was also, he discovered, angry at a fellow human being for her apparent neglect of her health. He felt heavily and negatively judged by her. Intertwined with his anger was his real concern that she might neglect her eye and lose the use of it. "Don't you realize what it means to lose your vision?" he wanted to shout at her.

The trainer than asked the resident to take her role once more, and began to interview her. Where did she live, he wanted to know. What did she do? What was her life like? How did she feel about her eye problem and about coming to this clinic ? About the previous treatment she had? How did she feel about this particular doctor?

Using all the information he had about the particular patient, as well as his awareness of how other members of her socio-economic, ethnic subculture thought and felt, the resident produced a picture of a young, black ghetto-dweller whose medical problems were only one more insult visited upon her by a life which was marked heavily by despair and a generalized feeling that there was little that could be done about anything which might alleviate any of her whole collection of emotional and physical hurts. In response to the trainer's questions and interviewing, she acknowledged that she did not believe that any white doctor would really give much a damn about her eye. If she had money or were white, than he would probably do something helpful for her. She reflected upon the ways she had been treated at clinics and at other helping agencies in the past. She expressed little faith in any of them or in the people that one met in them, especially anybody who pretended to have some genuine interest in her or her problems.

In the discussion which followed, the resident expressed a new-found awareness that he had been responding very personally to anger and resentments on the part of the patient which had likely originated in her past disappointments and rejections, and which did not belong in the current situation. Other residents in the group were able to identify similar experiences and similar feelings and reactions. Then the resident was given an opportunity to re-enact the situation, and was able to do so without experiencing the turmoil within himself that the original enactment had stirred up. The observers reported that his manner, while still highly professional, seemed a bit more personal and genuine.

Role training has at least three different applications in the field of psychiatry. It can, and has been, effectively incorporated into the treatment of patients, especially those individuals whose lack of common interpersonal skills may serve as a handicap to independent living outside an institution. Role training, in addition, provides a powerful and interesting media for fee teaching and supervision of group and individual psychotherapy. Finally, it is a unique and effective approach to the psychiatrist's function as a human relations consultant to his medical and para-medical colleagues of other specialties.

Role training is usually incorporated into treatment when the patient or client complains that there are certain specific situations in which he feels particularly inadequate or incompetent, or in which it is apparent to the therapist that the client is deficient in terms of adequate response. Variations of the critical situation may be presented to the patient, and he is encouraged to experiment with a variety of ways of responding to them, ranging all the way from his typical reactions to reactions which he would never allow himself. Other group members are allowed to demonstrate their particular methods of dealing with the problem area, and the patient is allowed to incorporate other styles and approaches into his own solutions. A few typical situations in which patients have been role trained for more effective performance include applying for a job, explaining one's psychiatric hospitalization to a friend, relative, or employer, behaving more freely with members of the opposite sex, avoiding stereotyped arguments with a spouse, etc. A further elaboration of role training in treatment will be found later in this paper in the discussion of the development of the role training method.

Although role training is very much a part of the Morenean system which has a rather complete psychotherapeutic approach of its own (psychodrama) role training can be utilized in the teaching and supervision of both group and individual psychotherapy, regardless of the theoretical orientation being practiced. Thus it is possible to teach Gestalt Therapy, Transactional Analysis, Psychoanalysis, or Behavior Therapy via role training, and to do so more effectively, often, than by conventional methods of teaching and supervision. In one kind of application of role training, the student therapist may be asked to reenact significant portions of a therapeutic session. Through the use of role reversal, soliloquy and other techniques, he has the opportunity to experience his interventions from the role of his patient, and to explore the roles and reactions of both himself and his patient in a novel and creative manner. This process adds new dimensions to the conventional techniques of supervision such as observation, tapes (video and audio), and process notes. It is a much more active process and it can lead to new speculations about the undisclosed motives operating in both patient and therapist. It is often helpful to the student therapist in his recognition and dealing with various aspects of counter-transference or other untoward emotional responses on his part. Role training allows the supervisor to correct technical errors on the spot in a context which closely simulates the actual therapy session. He also has the opportunity to demonstrate techniques and inter-ventions while the supervisee, in the role of the patient, has the opportunity to experience their effects. It provides a particularly effective way of dealing with the sometimes difficult problem of relating theory to practice.

The third use of role training in psychiatry has to do with the increasing emphasis upon the role of the psychiatrist as a specialist and consultant in human relations to the other medical specialities. More and more the psychiatrist is being called upon to serve as a teacher/consultant to his colleagues, assisting them in learning to assess emotional, behavioral, and social ramifications and manifestations of illness in their patients, and to respond to these factors as effectively as they have been taught to do with respect to the physiological factors of illness. His job is to leave non-psychiatric physicians more skilled in responding to the emotional needs of all their patients. Role training, placing emphasis upon performance and training rather than upon theory (often regarded by the non-psychiatric doctor as esoteric or even arcane) is a means of accomplishing this. It is a method which encourages the non-psychiatric physician to use his personal understanding of human behavior, derived in part from his experiences as a human being, in relating to his patients, and to do so without the necessity of translating their behavior into abstract theoretical terms.

An example of this use of role training can be found in the approach of a member of the psychiatric staff of a training hospital who conducted periodic in-service sessions for the nursing service. He sometimes began these teaching sessions by asking, Who's got the toughest patient in the House? A small hospital of about 100 beds, the nurses were usually able to agree upon whom was the biggest management problem of the moment, perhaps Mr. Smith on 2nd East. The staff from that floor would then be invited to show the rest of the group what kind of problems they were having with Mr. Smith.

One such Mr. Smith was perceived as an overbearing, demanding, grouchy, ungrateful, whining and generally unpleasant sort of person. The biggest complaint was his incessant (and to the floor staff, unnecessary) use of his call button. "You can hardly leave his room and

236

get back to the station before his light is on again," one nurse said, "And his room isn't that far from the station!"

Although he was in the hospital for a periodic assessment and stabilization of a serious heart condition, he was considered to be in an essentially non-critical condition. He was certainly not a candidate for intensive care at the moment, and his excessive use of his call button was not only felt by the floor staff as an uncalled upon irritant, but perceived by them as interfering with their responsibilities to other patients. They had tried a variety of tactics for dealing with Mr. Smith, all to no avail.

During the work which followed, all of the nurses who knew him and who had worked with him took turns playing the role of Mr. Smith. The trainer first of all had the nurses from his floor demonstrate his demanding behaviors. As they warmed up to their task, they delighted in doing so, and became more and more outrageous. The group enjoyed the humor of Mr. Smith's increasing demands and the nurses' eye-rolling responses of frustration to them.

Then the trainer began to interview him. A composite picture began to emerge of a rather severely frightened man behind a facade of demanding, whining gruffness. Although he had been told by both medical and nursing staff that his condition was considered satisfactory and that he was in no immediate danger of life-threatening distress, he also knew that he had a condition considered to be quite serious and which could become critical, as far as he knew, at almost any time. It turned out that no one had bothered to explain to him what to expect or when, and no one had taken the time to explore his feelings with him. Rather, most of the staff in direct contact with him had attempted to reassure him about how well he was doing. He could not accept this reassurance because to him it seemed like a denial of both the seriousness of his problem, and a discounting of his fear about it.

237

Pushing the call button could then be interpreted as an expression of both his fear and anger—fear that the staff might not respond quickly enough in an emergency, and anger that his real emotional needs were not being responded to adequately.

The consultant then asked the group to devise as many ways as they could for responding to Mr. Smith, assuming that the picture of him which they had created was approximately accurate. Various individuals demonstrated their solutions in further role playing and these were discussed at length. Finally, the staff from Mr. Smith's floor decided upon the strategy which they would try next. They first agreed that it might help to visit his room more often without being summoned. Then, realizing that reassurance was not providing the support that was intended, that they would train the nurses on the floor to simply listen carefully to any complaints or reports of his condition when he delivered them, and when it was appropriate, tell him that they would inform the doctor or the other floor staff. They also agreed that someone, his doctor most likely, should give him more information about his condition and what to expect.

A very moving moment then occurred when one of the nurses revealed that she had lived most of her life with a congenital heart condition which had been surgically corrected only when she was in her mid-40's . She corroborated the sense of fear and anxiety which the role players had produced in the role of Mr. Smith, and gave her endorsement to the new strategy which the floor staff had generated. She felt that the new approach incorporated a much better understanding of the patient's situation, as perceived by him, and could serve his needs as well as those of the nurses more effectively.

Reports of results from participants in sessions like the one just described are consistent with other experiences with the role training method. Frequently the solution arrived at in the role training session is

238

found to be very effective in the real-life situation. Sometimes it is not. In either case, participants usually report that the role training exercise helped to clarify matters, and even when the derived solution did not work, they were encouraged to continue experimenting with new ways of responding to the situation. Role training tends to train people to look at old situations from new and more productive perspectives.

Roles, Role Theory, Role-playing, and Role-taking

Role is a concept which has been the subject of considerable elaboration and much theorizing in sociology and social psychology for many years (Turner, 1956). For this discussion Role Theory as expounded by J. L. Moreno (Moreno, 1953) in his many theoretical and methodological contributions is most appropriate. Moreno's concept of role is a less abstract, more pragmatic notion than some, and recognizes that the term originally came from the theater where it refers to a character in a drama, represented by one of the actors. His role prescribes what the actor says and does, and especially how he relates to the other characters in the drama. Thus, Moreno defines role as the functional unit of behavior, meaning the functional unit of *social* or *interpersonal* behavior (Moreno, 1960, p 80.) Role is conceived of as having and exerting an organizational, identifying, and integrative influence upon behavior.

Role defines interpersonal interaction and implies a counter-role. For example, there can be no role of parent in the absence of the role of the child; no role of teacher without the role of student; doctor without patient, etc. Even the role of hermit requires others from whom to withdraw! From this point of view, all interpersonal interactions can be described and analyzed in terms of role relationship.

The role relationship between two people places demands and expectations upon both parties involved. Thus, if one individual assumes the role of teacher vis a vis another, the latter will feel some

239

kind pressure to assume the role of student, a demand which he may either accept or attempt to reject. Even if he accepts the complimentary role, the way in which he behaves within the role will, in turn, make some demands upon the teacher with respect to how the teacher behaves in the teacher role. There may be considerable adjusting on the parts of both people until each reaches some degree of comfort in the role relationship—or tensions will develop and both may tend to avoid or dread interactions with each other. If the relationship becomes quite comfortable and both meet the mutual expectations which they are making on one another, the teacher will tend to think of the other as a good student, rewarding him appropriately, while the student will tend to think of the other as a good teacher, and remember him fondly. These evaluations of each other may be somewhat independent of any objective measures either teaching or scholarship.

These dynamics probably play a much greater part in the direction and conduct of human behavior than is generally recognized and acknowledged. Much of what is experienced as personal satisfaction from social interaction reflects the skill and success we achieve in adequate and creative role enactment in concert with the significant other people in our social environment. The extent to which we influence and have an impact upon those around us is also a measure of role-taking skill.

Role has both collective and private components, its collective denominators and its individual differentials (Moreno, 1960, p 75). Thus, some roles, the roles of doctor, judge, teacher, and minister, for example, are heavily prescribed by social custom and convention, and often by legal definition. However, no matter how rigid these prescriptions are, considerable latitude is given to and expected from the individual who finds himself taking one of these roles. Hence, every individual who takes the role of doctor does so in a way which is different in some degree from every other doctor. However, if he behaves too differently from most other doctors, somebody, probably his patients, will wonder if he really is a doctor, no matter what diplomas hang on his walls.

The same doctor behaves a little differently with each patient—who behaves a little differently than every other patient. If this doesn't happen, the doctor may find that interactions between himself and his patients have become stereotyped, dull, and unrewarding—and the patient complains that the doctor doesn't really seem to be interested in him or his problems. Both may get the feeling of taking part in an assembly line process. The doctor may feel overworked much of the time, diagnose himself as a case of burn-out, and prescribe more frequent days or weeks away from his practice. The patient may look for another doctor who really seems to care. However, the infinite variation possible in the interpretation and taking of a professional role provides a tremendous opportunity for the creative expression of self, and is a tangible reflection of the individuality of the role-taker. Those who learn and utilize this principle tend to experience much less pressure and stress in their professional work, invest more energy and time in what they are doing, and achieve far greater satisfaction from it.

Members of any culture have considerable awareness of most of the roles which are to be found in that culture, and can actually portray roles which they have never been called upon to take in real life. Thus we usually find it easy to identify the roles that children are taking when they are playing house, or school, or cowboys and Indians. Sometimes we find it embarrassingly easy not only to recognize a parental role, but even to identify the role model This grasp of cultural roles is also reflected in the ability of most individuals to identify roles being taken or played by others, even when we have no first-hand knowledge of the role. It would take most people only a very brief period of time to recognize that they had encountered a robber, even if they had never ever had any previous experience with somebody in that role! It is also true, and can be demonstrated in role-testing experiments, that individuals generally have considerable capacity to produce accurate and valid role portrays of the common roles in the culture in which they live. This ability is usually far greater than the individual himself believes that he is capable of, and greater than could be expected from examination of his role history.

The learning of roles and role behavior is a complex process which usually involves several steps. First there is role perception and role - playing, both of which give shape to the role as it is being learned. These are followed by role taking, the actual enactment of the role behaviors in everyday life.

Role perception is the process of observing and identifying role be- havior being presented by somebody else. It occurs both intentionally and unintentionally. It is probably most effective in terms of influencing our subsequent behavior, when we are in the complimentary role and when it involves real-life situations. Thus, the major source of training-to-be-a-parent in our culture (and in most cultures, for that matter) comes from our experiences of being in the role of child, from which we perceive how our parents do their jobs of mothering and fathering. An interesting phenomenon sometimes results when the individual, finding some particular parental behavior to be especially repugnant vows that: "When I am a mother (father), I'll never do that to my child! "—only to discover him/herself years later reproducing the hated behavior, sometimes using the very words and tone of voice which the parent used.

A major source of role perception in professional medical training comes from observing the behavior of instructors and preceptors.

Thus, unregulated role modeling has become the primary method of role training now in use.

Some form of role-playing is the next step in the creation and development of a role. This can be systematic and elaborate, combining all levels of personality as occurs in formal, directed role training which incorporates perceptual, behavioral, emotional, fantasy and intellectual components into role enactment. Or, it can be no more than the instinctive rehearsal-in-fantasy which most of us engage in when we are facing an unfamiliar situation. Almost always, in the early stages of role development, there is at least a moment of role playing as the individual warms himself up to the actual role-taking situation. Repeated enactments at both role-playing and role-taking levels give rise to the individual role or role process which is than readily available for release and expression as life situations call for it.

Roles and role training have to do with the social skills as differentiated from the technical skills of a profession. It is possible for an individual without the technical skills of a physician to play that role or to trained to do so. A common example is the actor who plays the role of the doctor in a TV serial or in a movie. If he attempts to take that role, however, we call him an imposter and he is subject to serious legal sanctions. It is also possible for a person to develop the technical skills of a physician, perhaps to a very high degree, and yet do a poor job of taking the role. This state of affairs may result in patients who become anxious, angry, or who respond in other untoward ways. Neither this physician nor his patients will get much satisfaction out of their encounters, even though the doctor's diagnoses and treatment may be quite adequate as judged by objective standards. Usually these physicians find atypical positions in the profession in which the role behavior is not a problem. Sometimes they leave the profession.

Role training, role-playing, and role-taking should be distinguished from theatrical acting. While there are some distinct similarities between acting and role training, the goal of the former, ultimately, is to create the illusion of life as an artistic endeavor. Role-playing and role training are preparation for action in real life situations. They have to do with the roles in which we give expression to our very selves. Role training can be considered rehearsal for life.

The Role Training Process

Role training was originally developed by J. L. Moreno, M.D., in his work with a variety of socially deviate groups: juvenile delinquents, prisoners, and mental patients. Moreno recognized that members of these groups were vulnerable to institutionalization primarily because of a serious deficiency in social skills, or because the social behaviors which they had learned labeled them as members of a deviant population. Role training developed in conjunction with psychodrama as an approach to the problems of these groups, and both stem from earlier experiments in human creativity and role playing which Moreno had conducted in Vienna in the early 1920's (Moreno, 1947.)

Working with such groups as the delinquent girls of borderline intelligence who were the inmates of the Hudson School for Girls, Moreno had them practice, in role playing sessions, such common social roles as that of customers in a store or restaurant, employees seeking work, and a variety of other employee roles, etc. Out of such sessions developed the principles and techniques utilized in role training.

In the nearly 50 years since these beginnings, the role training method has been refined and elaborated into a highly effective, systematic approach to the teaching and learning of successful interpersonal behaviors. In addition to its use with the above mentioned populations, role training has been extensively employed in industrial settings where it is used for the training of managerial employees and others for whom a high level of skill in human relations is essential.

The mechanics of a role training session typically call for the identification and enactment of a situation which illustrates a common problem or difficulty which group members face with respect to the role under consideration. One or more trainees demonstrate and experiment with possible solutions to the problem situation. They are then usually allowed to experience and assess these solutions from the opposite role. Other solutions may be presented either by other group members or by special consultants or training staff who are present. The goal is not to decide upon the correct solution or response to the situation as much as it is to generate many potentially adequate and appropriate ways of addressing the situation. Members of the group may then be allowed to practice or rehearse courses of behavior which they want to master more thoroughly, or to face the situations which tend to arouse their anxiety most frequently.

Role training is probably most effective when it is combined with role taking activities, that is, when the trainee is actually involved in life situations which require him to take the role he is trying to master. In the medical school setting, for example, role training can be a helpful way for the student to prepare for his first encounters with patients. But even more effective utilization of the method may come when the advanced medical student, intern, or resident has the opportunity to bring to the role training session current problematic situations where he can examine them in the safety of the role training process. Alternative solutions can then be incorporated into his next interactions with patients.

Role training offers a number of advantages over the usual trial and error method through which the student physician commonly acquires his role behavior. There is the simple fact that the method provides a way for a means for the training center to establish some control over this important aspect of medical training. More efficient use of time accrues by separating the training of role skills from technical skills, and provides for greater safety. The role training process provides a place in which mistakes can be made without risking any serious consequences to the patient or the doctor-patient relationship. The role training situation also provides a way to get to the crux of the interpersonal aspects of this relationship in the absence of actual medical emergency.

The role training session allows for effective use of the collective experience of the group members who share with one another both actual experiences and potential solutions to common problems. It also offers a medium through which experienced physicians can show the trainees how they deal with a problem situation—an event which is more effective than telling them how it is done. In general, the action aspects of role training make it both more effective and more - interesting than discussion or other forms of teaching. Finally, the method stimulates the student's creativity in the expression of himself through his professional role, and thus, he learns a way which he can repeatedly utilize whenever he is confronted by special problems in human relations.

Despite the straight forwardness and apparent simplicity of the role training method, the importance of an experienced and skilled trainer must not be minimized. Successful role training requires a trainer who has been thoroughly trained in the spontaneity methods (sociometry, psychodrama, and role training), and an untrained leader can bring about seriously adverse results and negative learning.

Values, Ethics, and Role Training

The medical profession today is facing some new and unusual challenges, brought about, paradoxically, by the tremendous technological and methodological advances of the past several decades. Until quite recently, it was unheard of for a physician to have to consider that he might be going to far in his attempts to treat illness or preserve a patient's life. The very best he could do was never too much. Such is no longer the case, and more and more frequently the doctor must face decisions with respect to how much effort he should exert to maintain life. He now faces the prospect of confronting a family who may ask him to cease his efforts and allow a patient to die. His heroic efforts not only may not earn him the respect and gratitude they once did, but event the opposite may occur. These situations usually involve terribly complicated legal, ethical and philosophical issues, and the best of scientific thinking will not resolve them satisfactorily.

Ethical problems seldom involve knowing or not knowing right from wrong, but arise rather when one is faced with a situation in which two (or more) equally valid ethical principles are in conflict so that one must be chosen over the other. Abortion, for example, usually involves balancing one's responsibility to an unborn infant against responsibility to the mother who is carrying it. Severe brain involvement may confront the physician with the possibility that his best efforts may result only in a chronic brain invalid who will become a physical, emotional, and financial burden upon his family. The doctor who can deal with such infinitely complicated situations well, who can share the decision-making with the family involved without pushing too much of the responsibility onto them, or reserving too much of it to himself, and who can do so without risk of being overwhelmed by the stress which are inherent in such situations is certainly an exception rather than the rule.

247

For more than a decade now, there have been a number of attempts to confront and re-examine the meaning of death. Earnest Becker in his Pulitzer Prize-winning *Denial of Death* (Becker, 1973) has offered the interesting proposition that civilization and all of its achievements can be interpreted as immortality projects—attempts to deny and overcome the very fact of mortality. Whatever the validity of his thesis, he does document convincingly the widespread use of denial mechanisms which Man uses, individually and collectively, when confronted by his own mortality. Elisabeth Kübler-Ross Kübler-Ross, 1969) who is well-known for her efforts to demonstrate the operation of this factor within the medical profession, and who has clearly identified the adverse effects upon both patients and their families that the physician's discomfort when faced by the probability of death can have, has also shown that proper training can result in tremendously increased comfort for all involved— patients, family, and doctor.

As problems of this nature become increasingly prevalent (or perhaps it is more accurate to say increasingly apparent!) it becomes imperative that they be realistically addressed during the training of physicians. The major source of training in ethics and values probably lies in the rigorously honest and responsible behavior which medical colleagues demand from and offer to each other in the training setting, and from examples set by instructors, preceptors, and supervisors. And there is no reason for this condition to change. These experiences carry more weight than any formal presentation or informal discussion. However, role training can be utilized effectively in this kind of training, and offers some of the same kind of advantages that clinical training offers over classroom work. In the role training process, the trainee can be confronted realistically with a situation which requires him to make complex judgements and decisions—within the safety of simulation.

Summary

Role training provides a systematic procedure for increasing skills in communications and human relations, for exploring ethical values and attitudes. Some of its salient characteristics include the following:

1. Role training is action-oriented and emphasizes behavioral skills over knowledge. The trainee is taught to respond to the human being who is his patient rather than to analyze, interpret, or simply understand him.

2. Because it involves the trainee in action, the learning in role training is both visible and immediate. Corrections can be made on the spot.

3. Role training emphasizes autonomy and self learning. A feature of role training is that the student learns to respond effectively in unexpected and unpredictable situations.

4. Role training allows experimentation with a variety of different behaviors in a protected setting. One can focus attention upon the human factors in the doctor-patient relationship without jeopardizing a patient's well-being.

5. Although based upon the role theory of interpersonal interaction of J. L. Moreno, role training is essentially theory-free and does not require trainer or trainee to subscribe to any particular theory of human behavior. Indeed, trainer and trainees may hold to a variety of differing notions about why people are the way they are, and still work effectively together in role training.

REFERENCES

Becker, Earnest *The Denial of Death.* N.Y. : The Free Press, 1973

Kübler-Ross, Elisabeth. *On Death and Dying.* N. Y. : Macmillan Co., 1969

Moreno, J. L. *The Theater of Spontaneity.* N. Y. : Beacon House, Inc. , 1947

Moreno, J. L. *Who Shall Survive ?* N. Y. : Beacon House, Inc., 1953

Moreno, J. L. *The Sociometry Reader.* Glencoe, 111.: The Free Press, 1960

Turner, Ralph H. Role-taking, Role Standpoint, and Reference Group Behavior, *Amer. J. Soc.*, 1956, vol 61,   pp 316-328.

A BRIEF HISTORY OF PSYCHODRAMA AT TLC[7]

The Beginnings

Psychodrama has become such an integral aspect of TLC that a look at the history of the relationship between the TLC method and the psychodramatic method seems like a good idea.

The foundation of what has become known as the TLC method are the trial skills developed by Gerry Spence over his uniquely successful career as a trial lawyer. One component of those skills is related to Spence's involvement with the Human Potential Movement[1] of the 1950s, 1960's, and 1970's. Gerry was introduced to this movement by his great friend, John Johnson, who was the director for community mental health in Fremont County while Gerry was the county prosecutor. Wikopedia defines the Human Potential Movement as coming "out of the social and intellectual milieu of the 1960s... premised on the belief that through the development of human potential, humans can experience an exceptional quality of life filled with happiness, creativity, and fulfillment. A corollary belief is often that those who begin to unleash this potential will find their actions within society to be directed towards helping others release their potential. The belief is that the net effect of individuals cultivating their potential will bring about positive social change at large." And even though that is from Wikopedia, it is pretty accurate.

The movement made use of small group research and group psychotherapy techniques and principles to help individuals better understand both themselves as well as the dynamics of small groups. The aim was to achieve greater effectiveness, personally and

---

[7]Reprinted by permission of the Trial Lawyers College.
TLC here refers to Trial Lawyers College, a post-graduate training course for trial lawyers, founded by Gerry L. Spence in 1994.

interpersonally. One of the major players in this many-headed endeavor was the National Training Laboratories,[2] and this is where John Johnson steered Gerry Spence. Through his N.T.L. experiences Gerry learned much about individual and group dynamics, and this knowledge has been incorporated into his unique approach to trial lawyer skills.

Psychodrama and Trial Lawyers

In 1959, a law school professor, Howard Sacks, wrote an article on the use of psychodrama and role playing to improve the interpersonal skills of lawyers.[3] I ran across his article in 1997, while searching for some other references. I then discovered that Howard Sacks was the Dean Emeritus of the Connecticut Law School and lived about five miles from my house. He granted me an appointment to talk to him, but it was more a frustrating than a fulfilling experience. Although he recalled the article, he reminded me that it had been written a long time ago, and that he didn't remember much about the experiences which had gone into it. Nor did he seem to be particularly interested in hearing about the work we were doing at TLC.

His article relates details of a class he taught at Northwestern University School of Law, entitled "Professional Relations." He describes the use of role playing, one technique from psychodrama, but it is obvious that he did not actually use psychodrama or role training as we do at TLC.

Psychodrama came to TLC through the good offices of John Ackerman, a long time friend of Gerry Spence, with a major assist from John Johnson.[4] In 1975, Ackerman had become Dean of the National College for Criminal Defense (now the National Criminal Defense College.)[5] Up to that time, the NCCD classes had been all lecture. Ackerman introduced a modified NIT" style of teaching which had students on their feet and performing. While this was more effective than lecture, after a few years he was still not satisfied. Good lawyers developed their trial skills out of who they were, he observed. He wanted to find a way of helping young lawyers become more creative and spontaneous, using their intuition, dealing effectively with the ongoing unexpected events which always emerge in the courtroom. He called on John Johnson, a social worker, for help. "I don't know," John told Ackerman. "Let me think."

He finally reported back: "I think that psychodrama is what you are looking for."
In 1978, at the summer NCCD event, Ackerman, Johnson, and several others gave psychodrama a try, with Johnson directing after reading a book on psycho-drama. At that point, Ackerman says, they all saw the potential for psychodrama as a way for people to get in touch with themselves, to be real, open and honest. They also decided that it would be a good idea to find someone who was trained in using the method to do the directing.

Ackerman decided to offer a psychodrama program through the auspices of the NCCD. It was scheduled for the Fall of 1978, at Snow King Resort in Jackson, (Ackerman thought they needed a "remote location") and was billed as "The Criminal Trial: a Psychodramatic analysis." John Johnson found a psychodramatist, Kirsten Sonstegard, who had trained at St. Elizabeths Hospital in Washington, DC. She, in turn, asked Don Clarkson, who had been her trainer, to come along.

The event came off as planned, and the attendees were soon involved in intense personal psychodramas. "I saw the immense power of psychodrama," Ackerman reports, "and it scared the crap out of me." The feedback was that it was a fantastic experience, and the NCCD began offering a series of psychodramatic events. Don Clarkson was the major director. Gerry Spence attended the first presentation and was impressed with both the power of the method and with Clarkson as a director.

Psychodrama was not integrated into the skill training of NCCD. It was an "add on" program, Ackerman says with the sole purpose of helping people get out their intellectual selves into their emotional selves. It was seen as separate from training in trial skills.

In 1983, Ackerman left the college and interest in psychodrama left with him. The National Association of Criminal Defense Lawyers continued to sponsor some psychodrama events for a while, Ackerman says, but when the National Criminal Defense College was created to succeed the NCCD, psychodrama disappeared from the scene until TLC became active in 1994.

Psychodrama at the Trial Lawyers College

In February, 1994, I met a long time friend, Don Clarkson, at the Annual Conference of the American Group Psychotherapy Association. We had lunch together and Don asked me, "How would you like to do psychodrama with some lawyers at a trial lawyers college on a ranch in Wyoming in August?"

I had never done anything that sounded quite like that so I said, "Sure. Why not?" Then I asked "What is a trial lawyers college?"

"I don't really know," Don replied, "but a well-known lawyer, Gerry Spence—you probably know who he is—is doing one on his ranch in Wyoming. You've probably seen Gerry on TV programs like Larry King. He wears a buckskin jacket and has long white hair."

I didn't know who he was. I didn't watch Larry King or the other programs where Gerry was often a guest. Don mentioned the trials for which Gerry was famous. I, of course, had certainly known about the Silkwood, Marcos, and Randy Weaver cases, had followed them in the news, but the name of the lawyer had not imprinted itself in my memory. However, I was impressed.

"The college will last the whole month of August" Don informed me. "He wants us to do three days of psychodrama at the beginning and two days at the end."

Cut to July 31. An interminable ride on a 20 passenger shuttle bus with no shock absorbers and with lawyer shop talk ("What kind of law do you practice?" "Where you from?" "I had a case like that." "Civil and a little criminal." "Let me tell you about a case...") assailing our ears, ended near the cookhouse where Gerry Spence stood to greet the first participants in the first Trial Lawyers College.[6]

Don introduced me to Gerry who sent word for John Johnson to meet us in Gerry's cabin and took us there. "I want to fill you in on what we are up to and the part that you will play in this," he told us. After a typical Spence pause, "It all begins with me!" delivered in a stentorian Spence voice. I was a bit startled but soon realized that he was saying that a good trial lawyer's most important instrument was himself and that it was important for the lawyer to be in touch with the whole self, not just the intellectual apparatus. This, of course, was where psychodrama came in.

This idea reverberated strongly with me because I had been telling psychodrama students from the time I was Director of Training at the Moreno Institute, twenty years previously, "psychodramatist has two instruments, oneself and the psychodramatic method. To be a good psychodramatist, you must know them both thoroughly."

Spence went on: "trial is all about telling a story. You tell your client's story in voir dire, in opening statement, in examination and cross examination, and in closing.[7] Lawyers need to know how to tell the story."

Again my ears pricked up. I had also been telling students of psychodrama, "The job of the director is to help protagonists tell the story, the story of their lives or a story from their lives, to tell it through dramatic action." I thought to myself, "This Gerry Spence has some of the same kind of ideas that I do. I am interested in this and wish I could be more involved than just spending three days at the beginning and two at the end." Even then, without knowing anything more about the college, I had in mind the possibility that role training might somehow benefit this program.

I mentioned my interest in the college and what he was doing. Gerry invited me to stay as long as I wished but having not made arrangements to be away from my practice, I had to decline. I offered to spend some real time next year if the program was repeated.

I had developed a fairly high level of anxiety, wondering how lawyers would take to psychodrama. I could imagine that Clarkson and myself might be run off the ranch after the first day of psychodramatic experience, and the anxiety was not relieved by discovering that the students of this college had no idea of what they were in for. I ate dinner at the same table as Lynne Bratcher who asked me what kind of law I practiced. "I am not a lawyer," I replied. "I am one of your psychodrama-tists."

"Psychodramatist! What's that?" she demanded, obviously in considerable alarm.

"Oh, oh," I said to myself. "Nobody has told them about psychodrama." I evaded a direct answer and told her she would know by this time tomorrow.

I needn't have worried. For the most part, after the first shock, lawyers took to psychodrama more readily than the mental health professionals with whom I had been working most of my career. Don and I divided the participants into two groups. Since they included staff as well as students, the groups were large, about 30 each. Additionally both groups were being video taped. None the less, there was no lack of willing protagonists and eager auxiliaries. The dramas were quite powerful.

As arranged, Clarkson and I left on the fourth day of the college. I came back a day early, hoping to get a little feel for the college. When I showed up on the ranch, I was warmly, almost raucously welcomed back. The message I heard over and over was: It's been a wonderful month and it is all due to psychodrama! I was taken aback. I wasn't sure anybody would recognize me almost four weeks after I had left the ranch, especially considering the experiences which I assumed that they had had with each other. It was incredible to me, who was used to conducting one and two week psychodrama training workshops with 20 or 25 students, that three days (we only did two sessions a day) could possibly have that much impact upon a community of almost 60 people.

It would be two years before I could articulate what John Ackerman had intuited almost immediately upon experiencing psychodrama. On the bus and in their first contacts with each other, the participants[8] are not only getting acquainted but sizing each other up. Conditioned by law school and other educational experiences, they are wondering, perhaps only subconsciously, how they will stack up in the long run, who the competition for top honors is. Also, there is no shortage of fantasies, again unspoken, of astounding Spence or other faculty with a rousing opening statement or a devastating cross examination. Some may dream of hearing, "You are the best student who has ever come to this ranch," or "You are some kind of lawyer, son. I expect great things from you." Maybe, some might think, Gerry Spence will ask me to join his firm.

The first three days of psychodrama brings them to the realization that they are more than lawyers—they are human beings first. And each of them, all of them are very, very human. It reduces the competitive urges to a large degree, and to healthy competition in which one individual's win is everybody's win. Nobody has to lose. The only person I have to be better than is myself when I came. A cooperative and collaborative learning culture is thus established.

Don Clarkson and I conducted two days of psychodrama to close the first Trial Lawyers College.

The following year I spent three weeks on the ranch. Largely I was there, as far as I was concerned, to find out what went on at a trial lawyers college. These are some of the observations that I made:

6. There was a small staff that stayed on during the whole month:

7. There was a plenary group meeting every morning and whenever a new trial skill was introduced. Then the group broke down into six groups of eight participants for training sessions.

8. There were a series of guests, well-known, experienced lawyers, friends or colleagues of Gerry's, who came in off and on for varying periods of time, usually a few days. They mostly gave lectures or presentations or told court room stories. Some of them liked to teach. They did a lot of drinking in the evenings.

9. When they taught, each instructor taught what he/she had personally learned. There was certainly a lack of uniformity in what was being taught from one group to another. Only Gerry and Bob Rose were teaching Spence's approach.

10. The training followed what I have since learned was the NIT" or NCDC model. A participant was invited to conduct a voir dire or give an opening statement for ten minutes and then receive ten minutes of critique. It seemed important to give each student the same amount of time. In other words, participants were allowed to practice their mistakes for ten minutes and then get told what was wrong. I decided that my first hunch had been right: TLC could benefit from the Role Training model.

11. Instructors liked to talk a lot. Perhaps I could even say "lecture."

12. There were two instructors who varied from the common model: Bob Rose and Gerry Spence. Both intervened freely to help the participant improve his performance. They had invented their own approaches to role training.

I made a few contributions during the staff meetings which were held every afternoon, but mostly I was attempting to learn the culture and discover what the skills being taught were. I did direct a sociodrama of the Timothy McVeigh trial which had not yet taken place.

John Johnson and I also became good friends.

Graduates Come Back as Staff

Early on during the 1995 TLC, John Johnson suggested to Spence that we should bring graduates of the program back as staff members. Spence was not sure of the wisdom of this but eventually decided to give it a try. In May, 1996, eight graduates from the 1994 and 1995 classes, along with Gerry Spence, came to the ranch for a three-day training event in the use of psychodramatic techniques.

These people returned for the TLC session, and were pretty well regarded as "junior" staff, with the older, more experienced "guest" lawyers carrying the burden of teaching. This generated a certain number of problems between some of the "Old Bulls" and the "New Kids on the Block." The latter sometimes tried to introduce some of their newly learned techniques into the skill training sessions to the frustration and sometimes to the ire of the former.

One issue was becoming apparent. There was little consistency in what was being presented and taught in each of the six skill training sessions. Something different was being taught by each instructor. Spence and Rose were the only old staff teaching Spence's approach to trial skills.

An important event happened one day when Cindy Short and Ken Goldberg had a class to themselves. The trial skill involved was opening statement. The instructors had a student reverse roles with the client and do a simple re-enactment, setting the scene of the event at the center of the trial. At staff meeting, they reported that this intervention had a major impact upon the students' presentation. This was the first utilization of re-enactment in the skill training session. In turn, this led to the use of re-enactment during whole group demonstrations.

The use of graduate staff was sufficiently successful that an even larger group was invited to a three and a half day training session the following year. Role training models for some of the trial skills began to develop during this session and were put into practice during the TLC for 1997.

During TLC itself, something else happened. Responding to the ever-present questions of "What does psychodrama have to do with being a trial lawyer?" Spence began pointing out how psycho-drama techniques, especially role reversal, can be used as a means of preparing for trial. Participants were invited to discover their own connections between psycho-drama and the business of being a trial lawyer.

As time has gone on, fewer of the guest instructors of the early years of TLC have responded to invitations to TLC. The staff has become graduates and a few staunch supporters like Bill Trine, Paul Luvera, and Milton Grimes who have taken both to psychodrama and the methods which have evolved through TLC.

At the end of the first TLC, I was present when John Johnson, perhaps responding to remarks and questions from participants, asked Gerry what we were going to do about an alumni organization to maintain and build on the interest, energy, and the relationships which were established during August. Gerry's response was to the effect that if the group wanted an alumni association, they would have to organize it themselves. Tom Metier and Bill Trine have reported on the history of the F Warriors Club in a previous issue of *The Warrior?* Gerry had approved the use of the ranch for a reunion, provided that it was a working reunion (instead of a drinking one.)

John Johnson and I were invited to that first event and to subsequent ones. The folks attending shared with each other their experiences in trying what they had learned during TLC and there were some psychodrama sessions. No formal program was organized ahead of time and the first sessions involved generating the agenda for that reunion. This led to some complaints about the looseness or lack of organization from some of the participants.

The graduate staff members uniformly praised the staff training event, finding it extremely valuable in adding to and consolidating the skills and knowledge that they had picked up in TLC proper. I had often mused that it would be fantastic if all the graduates of TLC could come back as staff and have the same training experience that staff members got.

Of course that was impossible. And then the idea struck me that we could give the reunion attendees the same experience that we gave staff during the May training session. That idea was accepted and put into action and what started as TLC reunion became the Graduate Seminar. More recently, at the suggestion of Joane Garcia-Colson and with the encouragement of Gerry, there have been two weeks of Graduate Seminars with one week emphasizing psychodrama and the other emphasizing trial skills.

Development of Role Training Models

The development of role training methods specific to each of the trial skills has been and ongoing, evolving endeavor. In the first years of TLC, Gerry illustrated his maxim of "I'll show you mine if you show me yours," as a basic principle of Voir Dire by sharing the anxiety that he and most lawyers feel when they face a new panel of prospective jurors. "Even though I have stood up in front of a lot of juries like I am doing now," he might say, "I always feel anxious. I may not show my fear because I am good at hiding it. But I imagine some of you may be feeling some anxiety also." Getting some head nods, he continues, maybe pointing out that it is likely a new experience for many of the panel, that new experiences and the unknown often make us feel anxiety. And then, he might move on: "One of the things that I worry about is ___, and discusses one of his concerns about his case.

Without a greater elaboration of the process, this often resulted in students, in their skill training sessions, doing something like this:

"I'm anxious. I bet some of you are anxious, too?" Heads nod.

"Yeah. Well, so am I. I'm kinda scared. "Aren't some of you kinda scared?"

"Me, too." And the voir dire comes to a full halt, or worse, continues for a few more cycles which lead nowhere.

At the same time, with the assistance of Gerry and the trainers, a lot of people learned something that allowed them to move beyond this impasse. They indeed learned a new approach to voir dire. In the search for how this happens, Gerry and staff have realized that it is important for the lawyer not only to recognize one's own anxiety, but to understand where the anxiety comes from. It is not only the problems in the lawyer's case, but how the lawyer relates to those problems, and that involves exploring one's prior experiences and how those past events may be generating the present anxiety.

It was a student, Toni Bentham, in the second class of 2000 who first articulated the steps in the process that one goes through in presenting the lawyer's concerns to the panel in voir dire. Gerry immediately recognized what she had accomplished. Since then the process has been refined and is currently used as the basis for training TLC students in voir dire.

One example has to do with closing statement. In 1994, Gerry emphasized his belief that a trial is a matter of telling a story. "You tell the story in voir dire," he would say. "You tell the story in opening statement. You tell the story in direct examination, in cross examination. And you tell the story in closing statement."

In 1999, after a demonstration before the whole group, Gerry and I had a simultaneous insight. We compared notes afterward. "We haven't been doing closing right," Gerry told me. " You damn well better have gotten the story told by closing statement. What you need to do in the closing statement is to empower the jury to act in your client's behalf."

"I had the same reaction," I replied. "I hear people arguing about what they want the jury to feel when they go into the jury room to deliberate. I know what they want them to feel. They want them to feel ethical anger."

"I don't like that term, ethical anger. Got too many connotations."

"How about 'righteous indignation,'" I suggested.

"That's better. I like that."

Most recently Gerry and the staff have been looking at the way in which we teach cross examination, which Gerry has suggested is probably the most difficult trial skill both to practice and to teach.

Psychodrama and the Trial Lawyer

"Can a trial lawyer use the psychodramatic method?" is a perennial question at TLC events. The question arises because psychodrama has so often been defined as a method of psychotherapy. Lawyers, of course, should not engage in conducting psychotherapy. This writer along with others, has long advocated defining psychodrama as a method of communication. Psychotherapy is merely one of its many applications.

So the answer to the question about lawyers using psychodrama is "absolutely," provided that they know what they are doing and why they are doing it. There are three obvious applications of psychodramatic techniques for the trial lawyer. The first is to obtain information. This is designated "reenactment interviewing," following the lead of Nancy Drew who devised the term to describe her use of psychodramatic techniques in qualitative research.[10] Re-enactment interviewing conforms to a dictum I made in a previous issue of *The Warrior* to the effect that anybody who is entitled to ask a person "Tell me what happened," is also entitled to ask "Show me what happened." Re-enactment interviewing, making use of the action techniques of

263

psychodrama, always elicits more complete information than can be obtained through interviewing alone. Although we tend to think of memory as purely a neurological process, one quickly learns from psychodrama that memory is a neuromuscular or "whole body" event. Walking through an event always brings to mind a richer recall of the details.[11]

A second and absolutely noncontroversial use of psychodrama by the trial lawyer is its use in preparing for trial. This can be as simple as reversing roles with the significant participants in the trial, the client or clients, witnesses, the defendant, defense attorneys, judge, jurors, and anybody else who is involved. Some TLC graduates like to do a re-enactment of the events central to the trial in the role of the client or a primary witness. Others have effectively used action techniques in deposing the other side's witnesses.

A final application in the use of psychodrama is in the personal exploration of oneself. A basic feature of Gerry Spence's approach to trial lawyering is self-understanding and self-awareness. This includes particularly exploring one's emotional responses to the wide spectrum of experience. Many of the lawyers who have been most successful in applying the skills learned in TLC ascribe their success to the greater understanding of themselves which began with their introduction to psychodrama and continuing self-exploration afterwards. Jude Basile, for example, will tell you that psychodrama has informed and influenced everything he does professionally, beginning with his first contact with a prospective client.

Summary

So that's the story from one perspective. Psychodrama came to TLC and the ranch because of Gerry Spence and his appreciation of the power of the method in introducing ourselves to ourselves and learning to read and utilize our feelings instead of being immobilized by them. I feel very fortunate that I had the time and opportunity to contribute to the skill training functions of the College, and that Spence and the staff were so receptive to new ways of conducting training. I see my major contribution to TLC as having introduced the principles of role training to the College.

I am in awe of what lawyers have done with psychodrama and the ways they have invented to use the techniques of the method in trial preparation. The collective creativity of TLC lawyers is a tribute to the method and the spontaneous—creative philosophy upon which psychodrama is based.

Psychodrama has played an important role in Trial Lawyers College, but it is important to recognize just what its role is. It must always be kept in mind that TLC is the brain child of Gerry Spence, and that what has become known as the TLC method, is founded upon Spence's ideas of how to conduct a trial. By now a number of others, participants in TLC, have contributed to Gerry's insights, and Gerry, himself, has refined his understanding of how he does what he does.

The role of psychodrama has been to provide a vehicle to transmit the understanding and trial skills which Gerry originated. It is probably the ideal method for this purpose. Because it is founded on spontaneity and creativity, psycho-drama has helped keep the TLC program spontaneous and creative. The program continues to grow and develop. In every session, something new and original is discovered.

Originally introduced to prepare participants for learning, the role training aspect of psychodrama became the principal modality of skill training at TLC. As time went on, the value of the method in discovering and uncovering the story was realized. Finally came the insight that the techniques of psychodrama could be utilized by the trial lawyer in preparing for trial. Over the years there have been suggestions, complaints if you will, that there is too much psychodrama and too much emphasis upon psychodrama at TLC. Some have even expressed the thought that TLC has become more a school of psychodrama than of trial lawyer training. These ideas are quite misguided. Psychodrama is an instrument, one which is uniquely suited to Gerry Spence's method of being a trial lawyer. Some people learn to use it well; some don't. Practically everybody benefits from attending TLC.

ENDNOTES

1. Also known as "the encounter group movement," the human potential movement has its conceptual roots in existentialism and humanism and is a phenomenon of the social and intellectual milieu of the 1960s, formed to promote the cultivation of potential believed to be largely untapped in most people.

2. The Western Behavioral Sciences Institute was another of more creditable bases of the human potential movement, and one of its more famous staff members was Carl Rogers.

3. Sacks, Howard R. (1959). The use of psychodrama and role playing in improving the interpersonal skills of attorneys. *Group Psychotherapy*, 12, 240-249.

4. Ackerman had met Spence in 1966, while still a law student. He worked with Spence on the Joe Escobel case, the subject of Spence's book, *Murder and Madness*. Ackerman met Johnson through Spence.

5. Much of the material in this section comes from a taped interview of John Ackerman by Dana Cole in August, 1998, which Dana has generously shared with me. Some of the same material will be found in Cole's article, Psychodrama and the training of trial lawyers: Finding the story, published by the *Northern University Law Review*. Other ideas come from personal communications with John Ackerman, Don Clarkson, and John Johnson.

6. One of those participants shook Gerry's hand and fell on his knees, exclaiming, "I've shaken hands with Gerry Spence! That makes the whole trip worthwhile!" I remember thinking that Gerry Spence handled such obsequious homage quite well. In his place I might have sent the hapless student back where he came from.

7. That formulation has changed since 1998, when Gerry and I both had a simultaneous insight that the closing argument has a different function.

8. In his first welcoming speech in 1994, Gerry attempted to reduce the gap between participants and teachers. He rejected the terms "student" and "faculty." "We are all here to learn together and from each other. It's just that some of us are younger and less experienced than others, Young Warriors, while others are older, Old Warriors [pause] Old Fucking Warriors." Thus was born the designations, YW's and OFW's, which continued for some years. While those designations have gone by the board, the

notion remains of a laboratory in which what is taught and learned is a work in progress, contributed to by both student and faculty.

9. *The Warrior,* Spring, 2006.

10. Drew, N. (1993) Reenactment interviewing. Image: *Journal of Nursing Scholarship,* 25, 345-351

11. Another article on reenactment interviewing in is process.

REMEMBERING J. L. MORENO

Although I had met Dr. J. L. Moreno and Zerka a year and a half before, my initial visit to Beacon for 2 weeks of training in 1962 was really the beginning of a 12-year acquaintanceship, friendship, and studentship with Dr. Moreno, and the beginning of what has become a dedication and commitment to philosophy, theories, and methods that he originated. I have not been the same since that first visit—and for that, I am grateful.

Although I remember a great number of things about that first experience at the Moreno Institute, I think that the most impressive was the impact that his respect for his patients made on me.

The institute was a licensed sanitarium in those days, and a middle-aged woman was admitted for treatment on the same day that I arrived for training. The initial steps in her treatment were part of the 2 weeks of training that I was enlisted in. The trainees were the group for her first psychodramas. Her problem was that her former husband, from whom she had been divorced, would not let her go to live with the new husband to whom she had then been married. The fly in the ointment was that both the divorce and the marriage, to a young doctor whom she had not seen for several years, had been carried out by "radio waves." She was not very happy about being at the Moreno Institute. As a matter of fact, she was vociferously angry and upset, rather unpleasant to be around.

Now I had been educated and trained by very civilized people who would never intentionally be disrespectful to a patient or client, people who certainly taught me that it was important to have respect for one's patients. However, I had never seen the best of these listen to the delusional productions of a patient with the respect and concern with which Moreno listened to this woman. He made no attempt to challenge her reality as she presented it. He was not in the least skeptical about her story. He was interested in every detail, and when she discovered that he was not going to try to convince her that she was wrong, she warmed and blossomed and produced many details.

Then he promised to help her in every way he could to straighten out the unfortunate state of affairs into which her life had fallen.

In minutes, she was ready to do anything that he asked.

At this point, he introduced her to psychodrama and directed her in some scenes in which she simply enacted events in her daily life.

I had been taught to be polite. Moreno showed me what it meant to respect.

In the early years of my training, Moreno was still active in the training program, even though Zerka pretty well carried the brunt of the training responsibilities. He still came to the theater for some sessions. He directed psychodramas. He conducted didactic sessions. More often than not, the evening session was held in his living room, where he would talk with students about what they were learning and experiencing and use these interactions as a springboard for discussions that included theoretical and methodological issues as well as personal anecdotes.

I thoroughly enjoyed these sessions. I was quite in awe of Moreno and liked listening to the stories he told, even though I did not think that Moreno was above embellishing the facts from time to time. I did not always believe everything that he said about himself. So I have found it fascinating that Rene Marineau (1989) has authenticated so many of his stories and claims and discovered new ones. Of course, I fully believed the one about his having been born aboard a ship on the Black Sea, a "poetic truth," which both Gheorghe Bratescu (1975) and Marineau have identified as such.

Of many weeks that I spent in Beacon, one of the more memorable began on a Friday afternoon, November 22, 1963. As I got off the bus from Cooperstown, New York, news of the Kennedy assassination was just hitting the streets. This event, which altered the daily routine of everyone in the country, had the same effect upon our training group, possibly even more, considering the emotional climate of a training group at the Moreno Institute.

The training schedule did not hold up very well. We spent time watching TV, talking, and abreacting. But shortly after Lee Harvey Oswald, who seemed an unlikely assassin, was apprehended, Dr. Moreno decided that he could be of help, that if he could conduct a psychodrama with Oswald, he might be able to get at the truth. So he proposed that we move the whole workshop to Dallas for that purpose.

He immediately started contacting people he knew who might be able to arrange an invitation to Dallas. Before his efforts bore fruit, Oswald was himself assassinated, and this brought an end to the whole endeavor. To this day, I wonder what kind of experience I may have missed out on.

There is no scientific definition of genius, but there is no question in my mind that Moreno fits the description. I think that what makes one a genius is an act of perception. A genius is someone who looks and sees where others have looked and not seen. Newton reported that the concept of gravitation had come to him when he had seen apples fall from trees to the ground. What Newton saw was that the ground (the earth) fell up to the apple—a little bit. That is what no one else had seen; seeing that is what made Newton a genius.

I think that Moreno saw things that nobody else had seen. I think that at Mittendorf, Moreno *saw* black and red lines connecting people, just as Newton *saw* the earth move toward the apple falling from a tree. And as he watched these people trying to create a village from scratch, he began to understand the part that the forces of attraction and rejection played in the developing structure of the community.

He also *saw* spontaneity in the children in the gardens of Vienna and in his *Stegreif* players. It had a tangible, almost visible quality for him. And maybe he saw some other things, too, that he never quite articulated. A common experience for the genius is to regard his discovery as being "given" to him from outside himself. And I think that was true for Moreno.

Although a number of leading figures in psychiatry, sociology, psychology, and education recognized the importance of his discoveries, and William Alanson White, Adolf Meyer, Margaret Mead, Theodore Newcomb, Gordon Allport, Gardner Murphy, and Henry Murray, to mention a small sampling, promoted him and his work, he was not greatly appreciated by most of the rank-and-file of his psychiatric colleagues. He was considered a maverick and a troublemaker, probably beginning with his first attendance at an annual conference of the American Psychiatric Association in 1932, during which he demolished A. A. Brill's psychoanalytic critique of Abraham Lincoln. He was accused of calling attention to himself in ways that were not considered becoming or appropriate. I suspect that this perception was not altogether gratuitously bestowed upon him. Over the years since I met Dr. Moreno, I have met psychiatrists who tell me tales of Moreno's disrupting meetings at APA conventions with his outspoken criticisms of speakers or papers.

How accurate these stories are, I do not know. I can, however, easily believe that Moreno may have sometimes indulged himself in attempts to infuse a little spontaneity into the proceedings of a society that tends to function in an unbearably dull and boring manner and that his traditionally minded colleagues did, not appreciate his efforts. Instead, they tended to see him as offensively egotistical, grandiose, self-promoting, even megalomanic.

A recently published book, *Models of Group Therapy* (Shaefer & Galinski, 1989), includes a chapter on psychodrama and Moreno. The authors, who are generally quite positive about psychodrama, note that Moreno has had an impact on the whole field that has not been fully acknowledged, and they try to account for the fact that psychodrama and Moreno are not better known. They suggest:

Perhaps some explanation for this fact lies in three characteristics of Moreno and his approach. First is a self-acknowledged immodesty; undoubtedly his manner, his rather overbearing style, and his concern about being properly credited for his productions tended to put people off. (p. 101)

The other two reasons listed are (1) his writing style and (2) the fact that his thinking was too far ahead of the times.

In my opinion, a great deal more than egotism and immodesty was involved in Moreno's "self-promotion" and his insistence on recognition for his contributions. I believe that there were two factors that motivated some of the behaviors that may well have been self-defeating.

The first has to do with the fact that his thinking truly was novel. He did see things in unprecedented ways. It is a difficult matter to be that creative. Rollo May (1975) devoted a whole book to the topic of the courage to create. It does take courage to tell the world that you know things about it that nobody else has known before—and that the world badly needs to know. Moreno had tremendous confidence in himself. But I am convinced that, despite this great belief in his work, he still wanted, and desperately needed, confirmation from others that his ideas were indeed valid and that they made sense.

The other factor has to do with his conviction that he had seen some truths that were of extreme importance to humankind, and that they provided a key to the most pressing problem confronting humankind; namely, how do we avoid the pitfall of self-destruction? He had an answer—a spontaneous-creative social order—but he had trouble getting anybody to listen to him. He was a veritable Cassandra in that respect.

I think that he felt a tremendous responsibility. He had been permitted a glance deep into the nature of human society. He had seen a solution to the threat of self-destruction that we seem to find repeatedly confronting us. And he experienced a tremendous frustration in the difficulty that confronted him of generating enough interest to make that solution work. It is no easy thing to bring about a social revolution, a change in the social order, especially when the change Moreno had in mind required the collaboration of everybody!

His first attempts, of course, involved the *Stegreiftheater* in Vienna, and the Impromptu Theater in this country. It was only when neither seemed likely to achieve its purpose that he "retreated from the Theater of Spontaneity to the Therapeutic Theater" (J. L. Moreno, 1947) and to the development of psychodrama, combining the principles of spontaneity drama with his more traditional profession of psychiatry. Of course, it was here that he finally made an impact and achieved some measure of attention.

Despite the fact that he credits his retreat to psychodrama with keeping his work alive, it was not accomplished without some cost. If you wanted to start a worldwide revolution, the traditional mental health professions were not the most likely comrades. Historically, members of these professions can more frequently be accused of establishment bias than of revolutionary radicalism.

What happened, of course, has been documented by Zerka Moreno (1969) in her paper, "Moreneans: Heretics of Yesterday Are the Orthodoxy of Today." As Moreno created methods, concepts, and techniques, all based on a philosophy of a world system and designed to move society toward a spontaneous-creative social order, colleagues appropriated them and put them to use in ways that only perpetuate the status quo of a technological, legalistic, conserve-conscious social order. As Zerka puts it, the ideas and concepts have been separated from the parent philosophy and from the long-term goal—a world order that can bring peace.

Perhaps, looking at things from this perspective, it makes more sense that he was sensitive about issues of priority and attribution and that he lamented that his instruments for social change had been borrowed, often without acknowledgment, and used for other purposes, whereas his theoretical ideas, the mother lode from which the instruments came, "gathered dust on library shelves" (J. L. Moreno, 1953).

There is still danger, it seems to me, that the success of psychodrama as a therapeutic modality can stand in the way of the development of some of its broader uses, in education, for example, and that this dynamic can delay the establishment of a spontaneous-creative social order. We reduce psychodrama when we think of it simply as a method of psychotherapy. It is of a much broader scope than that, and if we include role training, sociodrama, and spontaneity theater, we expand the applications even further.

Perhaps Moreno was overbearing, even immodest. I guess that a man who spends 10 years of his young adulthood taking upon himself the role of God—and then talks about it—either has an overabundance of ego or a pathological lack of it. Moreno certainly did not suffer from the latter, despite the claims of some of his detractors.

It is easy for people to misunderstand Moreno's God-playing, to take it as evidence of overbearing ego. It is not always understood as the very serious endeavor that he embarked upon, one in which he assigned himself what he would later refer to as a delusion. He said to himself something like: "What if I really am God? What if I have created this world, and I have created myself as a man in this world? This world is obviously not like I want it and it is incomplete. Now what do I do next? How do I go about making it better?" Thinking of this ilk does not engender in one a feeling of egotism. It gives one, rather, a feeling of overwhelming responsibility and a very serious need to get busy making things better. I think Moreno felt that responsibility for the rest of his life. He encouraged other people to experience it also. And it was from this experience that his ideas of spontaneity-creativity and God arose.

If he was egotistical, he could also be quite humble. From Zerka, I have heard the story of a patient who became psychotic after giving birth to a child. She recovered under his care, but he told her when she left his sanitarium, "I don't think you should have any more children. I am afraid that you could have another psychotic episode if you do." Some time later, she was admitted to his sanitarium. She had had a psychotic break following childbirth. "Well, Doctor," she said to him, "I guess you can say 'I told you so.' " Tears filled his eyes. "My friend," he replied, "I'd give anything to have been wrong."

Another story from Zerka: The Moreno family is doing some last-minute Christmas shopping on the 24th of December. They are in Macy's, Dr. Moreno in his usual black suit and bow tie. A very frantic lady rushes up to him and says, "It's just terrible. It's going to ruin our whole Christmas, and you've got to do something about it. The sofa came and it has a terrible rip in the fabric." It was pretty obvious what had happened and that she had mistaken Moreno for an employee of the store.

"Do you have your sales receipt?" he asked her. She did. He took it. "Wait right here," he told her. "Don't move and I'll be right back."

A few minutes later, he returned with a floorwalker, introduced him to the customer with a flourish, and said, "Here, Madam. This gentleman will take care of everything for you." She may have thought that she had been taken care of by Mr. Macy, himself!

And this story comes from Ann Quinn, nurse and residence manager at the Moreno Institute. During his final illness, Quinnie took care of him. Every day, about noon, she would go down to the house, give him a bath, and do whatever she could to make him comfortable. Just a week or so before his death, she walked in one day and he seemed rather depressed. "Is this any way for the great Dr. Moreno to be acting?" she said in an effort to cheer him up.

"Miss Quinn," he replied in a sad little voice, "I wouldn't recommend myself to anybody today."

Moreno's priorities with respect to psychodrama and sociometry, group psychotherapy, marriage and family therapy, and the influence that he has had upon the present-day practice of psychotherapy, both individual and group, are already well documented and probably challenged by nobody today. At the same time, there are really only a handful of us who make use of his most potent methods, psychodrama and sociometry.

The questions that almost everybody who experiences the excitement and power of psychodrama asks are: Why haven't psychodrama and the other contributions of J. L. Moreno caught on? Why haven't they received a wider hearing and greater application?

We can point to his personality (Moreno himself did), to the fact that he published his own works, depriving them of the potential for distribution that an established publisher might have been able to give them, his writing style that was admittedly difficult, and to the fact that his work was so far ahead of his time. It was not only ahead of his time; in many ways, it is still ahead of *our* time!

And yet, I think that the real answer to why there are only about 300 certified psychodramatists/sociometrists in the world is that these methods are scary and, in some respects, even potentially dangerous. This notion was brought home to me rather pointedly at this conference by the presentation of Dr. Robert Blake (1989), who discussed the use of sociometric methods by law enforcement agencies in attempts to combat drug distribution networks (see p. 148). Dr. Blake pointed to the ethical question of using sociometric methods to destroy an organization. Although few would quibble about the attack on drug-dealing rings, the situation implies that someone with a significant amount of sociometric information about an organization, a business, for example, could use it for destructive as well as for productive, creative purposes.

Psychodrama, too, scares people—sometimes the participants and sometimes the administrative people of the agency in which it is being used. Obviously, a majority of the therapists in this country are quite willing to do without its unique potency in the field of psychotherapy, and I reluctantly propose that fear is one of the reasons.

Nor has Morenean theory advanced significantly beyond the point at which Moreno left it. As a matter of fact, applications of Morenean methods have been so sparse that there has not been the kind of interplay between theory and technology that is probably necessary for the development and advancement of both.

With a few notable exceptions, most applications of Morenean methods and theory are in one field, mental health, and almost nobody identifies actively with Moreno's original goal of bringing about a spontaneous-creative social order.

Why is that so? Was his idea of a spontaneous-creative social order just one man's crotchet? Do we no longer need it? Or have we already got it?

I think that if J. L. Moreno could be here in more than spirit, he would be extremely excited about the current major outbreak of spontaneity in the world. I refer to *glasnost* and *perestroika*, of course. And I think he would say something like, "If only Gorbachev had a sociometrist!" I don't have any doubt that he would have long ago been on the phone, trying to call the Kremlin to offer both his advice and help.

I think that he would agree, however, that the spontaneous-creative social order is still a long way off.

In closing, I want to share with you another week that I spent at the Moreno Institute. It was in May 1974, exactly 15 years ago. I was between quarters at the university where I was a faculty member. Moreno was on his deathbed.

When Miss Quinn saw me coming into the student quarters, she said, "I'm so glad you are here. I need help with Doctor when I bathe him." She explained that he always shifted way down in his bed, and she had to have help to get him moved up where he belonged. She would ask one of the students in residence to help her, but, she said, she knew that Dr. Moreno did not like to be seen in his current condition by someone whom he didn't know. It would be better now that I could help.

And so, every day, just after lunch, Quinnie would call, and I would go down to the house to help her. The first time I went into his bedroom, Quinnie said, "Look who's here, Doctor." Moreno opened his eyes and when he saw me made an old familiar gesture. Lying flat on his back, his arms reached straight out to me. And he smiled.

And then, on Wednesday of that week, Quinnie called me as usual. Zerka was taking some sun on the patio. She had been with him all morning. We spoke briefly. The end was near, she told me. I went in. Moreno was not very responsive, and, as we were moving him up in his bed, Quinnie stopped and said, "Get Zerka." I did. Moments later, he was dead.

"Psychodrama is modeled after life," Moreno liked to say. He also said, "There is no death in psychodrama."

And that is one way in which life and psychodrama are different. Life comes to an end, at least for us as individuals, and that end is death.

Maybe Moreno was not recommending himself to anybody that day toward the end, when Quinnie talked with him.

But I recommend him.

I recommend J. L. Moreno to everybody.

REFERENCES

Blake, R. (1989). *Ten major opportunities for sociometric technology in corporate achievement.* Paper presented at the 47th Annual Meeting of the American Society of Group Psychotherapy and Psychodrama, New York, NY.

Bratescu, G. (1975). The date and birthplace of J. L. Moreno. *Group Psychotherapy and Psychodrama, 27,* 2-4.

Marineau, R. (1989). *J. L. Moreno: Poetic and psychological truth vs. historical truth.* Paper presented at the 47th Annual Meeting of the American Society of Group Psychotherapy and Psychodrama, New York, NY.

May, R. (1975). *The courage to create.* NY: Norton.

Moreno, J. (1947). *The theater of spontaneity.* Beacon, NY: Beacon House.

Moreno, J. (1953). Preludes of the sociometric movement, in *Who shall survive?* Beacon, NY: Beacon House.

Moreno, Z.. (1969). Moreneans: The heretics of yesterday are the orthodoxy of today. *Group Psychotherapy, 22,* 1-6.

Shaefer, J. & Galinski, M. (1989). *Models of group therapy* (2nd ed.). Englewood Cliffs, NJ: Prentice Hall.

ABOUT THE AUTHOR

John Nolte earned his doctorate in clinical psychology from Washington University in St. Louis in 1958. Introduced to psychodrama by Leon Fine in 1959, his erstwhile fellow graduate student, John was impressed by the power of the method and soon began training at the Moreno Institute with J. L. and Zerka T. Moreno. Most of his career has been in the mental health field, and he has held positions in the Departments of Mental Health of both Missouri and Illinois where he was the director of the first clinical unit of Adolf Meyer Zone Center in Decatur. He then went on to become a charter member of the faculty of Sangamon State University in Springfield, Illinois, a position he left in 1975, after the death of J. L. Moreno, to become Director of Training for the Moreno Institute. John estimates that he has directed, supervised, or otherwise participated in over 4000 psychodrdama sessions.

In 1972, he founded the Midwest Center for Psychodrama and Sociometry, [www.nationalpsychodramatrainingcenter.com] which has since become The National Psychodrama Training Center, conducting professional psychodrama training workshops in all regions of the country. Although a mental health professional, John has always been an advocate of non-clinical uses of the psychodramatic method. In 1994, he had the opportunity to develop applications of psychodrama in the world of trial lawyers when he was invited to join the staff of the Trial Lawyers College, founded by Gerry Spence, one of America's foremost trial lawyers. An account of that endeavor is included in this book.

John lives in Hartford, CT, with his wife, Nancy Drew, and welcomes correspondence by email: PDPapers.nolte@gmail.com.